规划师眼中的荷兰小镇
Dutch Towns as seen through planner's eyes

戴军　Joost van den Hoek　著

李露　傅蓁　陈辰　编译

中国林业出版社
China Forestry Publishing House

图书在版编目（CIP）数据

规划师眼中的荷兰小镇 / 戴军，（荷）约斯特·范·霍克著；
李露，傅蓁，陈辰编译. —— 北京：中国林业出版社，2017.4

ISBN 978-7-5038-8215-9

Ⅰ. ①规… Ⅱ. ①戴… ②约… ③李… ④傅… ⑤陈…
Ⅲ. ①小城镇－城市规划－荷兰 Ⅳ. ① TU984.563

中国版本图书馆 CIP 数据核字 (2017) 第 063158 号

规划师眼中的荷兰小镇

中国林业出版社·建筑分社
责任编辑：纪 亮 陈 惠

出版：中国林业出版社（100009 北京西城区德内大街刘海胡同 7 号）
网站：lycb.forestry.gov.cn
印刷：北京利丰雅高长城印刷有限公司
发行：中国林业出版社
电话：（010）8314 3614
版次：2017 年 4 月第 1 版
印次：2017 年 4 月第 1 次
开本：889mm×1194mm 1/20
印张：10.5
字数：300 千字
定价：98.00 元

前言 INTRODUCTION

通过荷兰特色小镇案例，启发当代中国特色小镇的规划与设计。

本书核心

什么造就一个特色小镇？又是什么造就荷兰特色小镇？这两个问题是本书的核心。

特色小镇的设计，在未来几年，将是政府面临的诸多主要挑战之一。因此，本书通过分析不同类型的荷兰著名特色小镇，为中国特色小镇的设计提供灵感。

同时，笔者也希望这本书对那些计划访问荷兰的城市设计专业人士提供指导。

小国家：全球化

荷兰是欧洲三角洲地区沿北海的一个小国家。然而，荷兰经济和文化的足迹，遍布全球，远远超过人们基于其规模小、人口少的预期。荷兰的财富和全球化有着悠久的历史。

所有历史学家一致认为：荷兰的特征和文化与其早期高度发达的小镇网络密切相关。这些小镇主要通过可航行的水道连接。也正是这些16世纪形成的小镇网络，奠定了荷兰昔日经济辉煌和如今生活高品质的基础。

规模适中

与中国的城市和城市群相比，荷兰主要城市的网络和上海的规模尺寸相当。

首都阿姆斯特丹只有77.9万居民，鹿特丹和海牙也仅有50多万居民。在一个1800万人口的国家，荷兰只有30个城市和小镇，有10万多居民。然而，人们在每个小镇都感受到生活和工作的完美平衡、城市经济的多元化，体验耕种休闲设施和文化旅游。这些元素都与小镇景观紧密联系。

荷兰小镇

几乎所有的荷兰小镇都可以追溯到镇中心数百年前没有汽车、飞机和电脑的时代。历史从未远去，从荷兰人的生活环境依然能看到他们祖先的印迹。这些历史悠久的市中心尺度宜人，仍然是非常受欢迎的生活、工作和休闲区域。

当然，荷兰小镇的历史中心也得到了扩展和更新。与水基础设施、汽车基础设施、工业化、公共空间和建筑物有关的现代化和城市扩张，占据主导地位。相反，曾经消失的街道和运河又重新恢复了对当代生活的装饰意义，成为全球化经济玩家的栖息地。

社会和经济发展的产物

荷兰小镇以其清晰、明确的结构而闻名。然而，很难将荷兰小镇的形状和规模纯粹定义为规划和设计的产品。人们需要对小镇的社会、经济历史及重大活动有基本把握，以了解小镇的演变。大多数小镇最初的发展都与当地的土壤、水网和内陆有关。

随着历史的变迁，小镇发展成为我们现在看到的城市群落：既拥有历史结构，又增加了生活、工作、基础设施和娱乐等功能。它们是在公共框架内、大多数由私人开发的分层结构。

历史再现

有些小镇已经大大扩展为城市，活跃在世界经济的舞台。而有些小镇则保持了原有的尺度，但或许也失去其全盛时期的地位。

无论哪种小镇，都保留了历史的痕迹，如今以建筑物、街道、港口、河流和水系的形式展现在世人面前。多数荷兰小镇因其历史背景、历史价值和历史遗迹，吸引了大量游客，休闲旅游已成为经济的一个组成部分。

关注特色

21世纪城市发展的趋势之一是，经济和文化全球化下形成的重要节点大城市与较小规模、更关注局部的城镇之间的割裂。在荷兰，这个趋势和在中国一样明显，因为具有国际导向，发展速度也比二三线城市快。

本书不会专注于规模较大、发展完善的城市，而是关注荷兰的特色小镇。这些小镇各自鲜明的特征，保持了小镇的生命力；也为其同中国特色小镇的对比，提供了借鉴。

聚焦历史棱镜

为了获得灵感，了解当代的荷兰小镇，笔者将通过历史的镜头来审视它们，从经济活动、城市规划和建筑亮点等角度分析它们。

本书开篇简要介绍了荷兰黄金百年的小镇情况，及其随后的扩展和20世纪后的工业化时期。书中还概述了荷兰小镇在21世纪将面临的挑战。本书主要分析了5种不同类型下的15个荷兰特色小镇的案例。

目录 CONTENTS

前言　INTRODUCTION

1 / 经典荷兰小镇和黄金世纪　THE CLASSIC DUTCH TOWNS AND THE GOLDEN CENTURY

4 / 工业化、现代化和近郊化　INDUSTRIALIZATION, MODERNIZATION AND SUBURBANIZATION

10 / 当代荷兰城市的挑战　CHALLENGES FOR THE CONTEMPORY DUTCH TOWNS

12 / 特色小镇分类　FEATURE TOWN TYPOLOGIES

17 / 历史小镇　HISTORY TOWNS

45 / 水小镇　WATER TOWNS

75 / 艺术小镇　ART TOWNS

93 / 产业小镇　PRODUCT TOWNS

133 / 科技小镇　SCIENCE TOWNS

203 / 后记　POSTSCRIPT

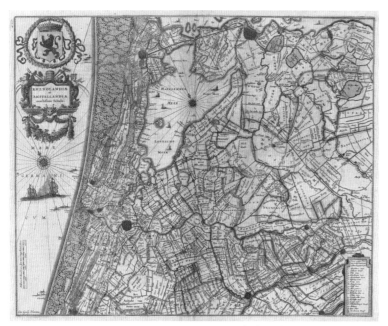

1640年左右荷兰中央地区地图：
由可航行的水路连接的水镇网络
*Map of central holland around 1640:
A network of water towns connected by navigale waterways*

1581年多德雷赫特地图：
荷兰早期的重要贸易城镇
*Map of Dordrecht 1581:
Important early Dutch trade town*

荷兰荷兰东印度公司商船：
荷兰强大的航海实力
*Dutch merchant fleet of VOC:
Seafaring Dutch powers*

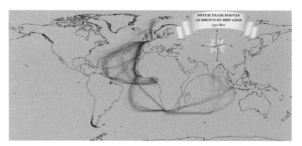

16～17世纪荷兰东印度公司全球贸易路线：
与美洲和亚洲的贸易
*Global trading routes of the VOC during 16th～17th century:
Trade with the Americas and Asia*

经典荷兰小镇和黄金世纪
THE CLASSIC DUTCH TOWN AND THE GOLDEN CENTURY

平地上的特殊城镇网络

中世纪后,在荷兰时期,欧洲莱茵河和马斯河河口形成了强大的小镇网络。三角洲的沼泽平地容纳了大大小小错综复杂的水系。借助通航水道互联互通的荷兰小镇,对荷兰的早期贸易发展起到了至关重要的作用,由此形成了各个贸易小镇。

给养城市

荷兰小镇兴起的前提之一是可靠的农业系统,促进了商业生产、食品贸易和规模产品,超越了农村自给自足的形式。荷兰小镇的崛起离不开小镇与周边区域的联系。不同种类的食物和产品从周围的村庄,通过船只被运送到附近的小镇和周边地区。

早期城镇化

荷兰小镇在国防、行政管理、水管理、贸易、农业和宗教信仰等方面有共同的特点。然而,通常小镇的发展与当地背景、基础设施和居民技能相关,从而形成独特的产品和服务特征,留存至今。这个时期的荷兰是世界上城镇化最高的地区,几乎有50%的人生活在城市。

黄金世纪

在16世纪与西班牙的80年战争中,荷兰的城镇和各省聚集起来,形成了荷兰共和国。在荷兰西部,形成了一个强有力的城市群,就是如今大家熟知的阿姆斯特丹、莱顿、哈勒姆、鹿特丹、代尔夫特、多德雷赫特。荷兰共和国转变为一个无与伦比的贸易、交通和水上战争平台。这一时期使荷兰成为了世界上最富有、最有文化、最强大的国家。因此,17世纪被称为荷兰的"黄金世纪"。

全球贸易

荷兰的航海水乡网络体系使国人的生产技能、贸易知识和航海能力达到世界水平。这一转型的主要加速器是荷兰联合东印度公司。该公司是由政府成立的,经营荷兰和亚洲之间的贸易。它不仅是世界上第一家跨国公司,也是世界上第一家公开上市的公司。早期的金融创新也促进了荷兰帝国的崛起。

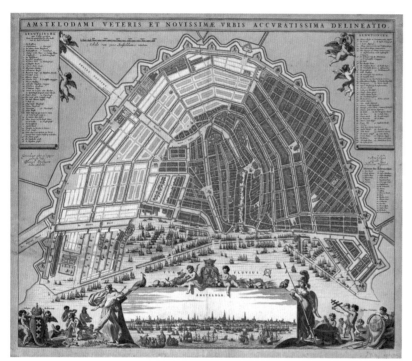

1662年阿姆斯特丹地图：
建设中的有运河的商贾城镇
Map of Amsterdam 1662:
Merchant town with canals under construction

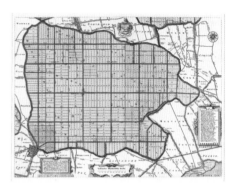

1658年比姆斯特尔圩田：
人造景观
Beemster polder 1658:
Manmade landscape

1685年阿姆斯特丹运河：
黄金世纪的建筑
Amsterdam canal 1685:
Buidings in the golden century

1693年阿姆斯特丹市政厅：
公民社会的开始
Amsterdam Townhall 1693:
Beginning of the civic society

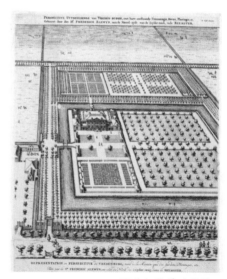

17世纪的景观设计：
文艺复兴的美学理想
Landscape design of the 17th century:
Renaissance ideals of aesthetics

经典荷兰小镇和黄金世纪
THE CLASSIC DUTCH TOWN AND THE GOLDEN CENTURY

"人造"国家

由于荷兰大部分区域处在海平面以下，荷兰人都说他们的土地是从海洋借来的。为了在这种情况下生存与发展，一套精细的水管理技术与诸如堤防、运河、风车这样的事物诞生了。操作这一系统，需要城镇、政府、农民和地主之间的高度协调与配合。许多历史学家认为，水管理派生出的集体组织技能是荷兰帝国迅速崛起的另一个重要前提。

商贾小镇

在黄金世纪，贸易主导的水城达到鼎盛时期。阿姆斯特丹和大运河就是最著名的例子。水上交通与城市设计的融合是个例外，至今还是。货物运输的船只可以停靠在商贾房屋前，这些房屋作为储藏室、陈列室、办公生活区和家庭住宅，商业水道沿线的商贾住房促进了全球的贸易联系。这种城市设计与水道连通的方法，形成了创造贸易机会的网络体系，好比21世纪的互联网。

景观探索

海外贸易赚取的利润，再投资到荷兰小镇。运河周围的小镇扩展投入了大量资金，随之产生的另一个影响是周围景观的殖民化。17世纪，富裕的阿姆斯特丹商人投资了第一个大型圩田区：阿姆斯特丹北部贝姆斯特。圩田区的土地租给农民，供给荷兰食物。近70公里、采用文艺复兴设计理念的圩田区，展示了早期惊人的工程技术特点。这种人造景观的美，闻所未闻。如今，贝姆斯特已被列入联合国教科文组织世界遗产。

民间社团与象征

在那个时期，其他欧洲国家的组织权力是由君主和教会赋予的。荷兰小镇的崛起非同寻常，因为它是由人民群众组织的，后来被称为民间社团。正是由于这个原因，荷兰城市的主要建筑不是统治者的宫殿，而是非常低调、控制造价的市政厅、议会大厦、大学、剧院和植物园。

适度、高效和功能

教会在荷兰城市处于一个相对次要的角色，因为宗教信仰过去是个新事物。教会建筑在荷兰小镇随处可见，但从来不是组织的主导力量。事实上，荷兰由荷兰人民统治，这也解释了为什么大尺度的、表达政府君主制、宗教绝对权力的城市元素，如广场和宫殿，在荷兰水乡城镇的布局和轮廓中是不存在的。在制图师Blaue绘制的著名地图中，小镇似乎是一个高效的贸易港口，点缀分布着小街道、小市场、特别的房屋和尺度适中的公共设施。最难以捉摸的财富标志可以从富贾住宅的门面上找到。

停滞

荷兰的黄金时代从1580年左右一直持续到1700年。尽管荷兰一直是重要的贸易国家，1700年后经济遭受了一段时间的衰退。这主要是因为其他作为国际贸易力量的欧洲国家的崛起。英国成为继荷兰之后具有世界影响力的主要贸易国家。在法国统治之下，荷兰沦陷。荷兰的小镇，除了在黄金世纪和荷兰共和国时期，再无显著发展。

1868年荷兰铁路：工业化前奏
*Dutch railways 1868:
Prelude to industrialisation*

1880年左右的鹿特丹：港口城镇的起点
Rotterdam around 1880: Start of harbortown

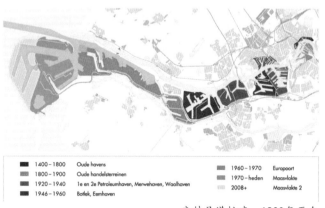

鹿特丹港扩建：1800年至今
*Expansion of Rotterdam Harbour:
From 1800 until now*

1890年左右的鹿特丹港：工业港口
Rotterdam harbour around 1890: Industrial harbor

1895年左右的鹿特丹韦斯特辛格：改善卫生城市条件
*Rotterdam westersingel around 1895:
Improvement of hygienic city conditions*

工业化、现代化和近郊化
INDUSTRIALISATION MODERNIZATION AND SUBURBANISATION

工业化的缓慢开始

荷兰的19世纪很多时候被描述为"冬眠时期"。这一时期，荷兰从黄金世纪的繁荣中逐渐恢复。荷兰的工业化进程起步相对较晚，规模较小。然而，现代化进程开始了，如铁路网的引入以及大量的运河和土地项目。新焦点聚集到城墙外围的城市发展和火车站周围。到19世纪末，大多数荷兰城市拆除防御墙和防御工事。关于现代和卫生城市规划的新想法被应用，实现了雨水净化处理和污水系统管理，保证了大多数城市人口的健康生活。

鹿特丹扩张

19世纪对于鹿特丹的意义，犹如黄金世纪对于阿姆斯特丹的意义。工业化的早期效应带来了鹿特丹港口的大幅扩张，使其成为世界上最大的港口之一，拥有与欧洲腹地连接的强大的铁路和公路系统。工业化的规模和不断发展的技术在马斯河南侧大型码头设施、吊装操作和货物存储中的应用变得明显。同时，随着"辛格（运河）计划"的推出，鹿特丹在卫生城市发展的形式中处于领先地位。直到20世纪初，荷兰城市由于缺乏下水道和开放的雨水处理系统，空气中总是弥漫着不健康的臭气。"辛格（运河）计划"为拥挤的旧城区和新地块提出了一套结合绿色空间的水处理系统，提供了高低端混合的住宅和商业设施。今天，"辛格（运河）计划"之于鹿特丹，就像大运河之于阿姆斯特丹。

系统的城市规划介绍

在19世纪末，城市规划和设计已成为一个独立的学科和专业。到目前为止，城市规划是国防专家、水处理工程师及建筑师关注的领域，而城市的建设是基于一个随意的维度基础。大多数房屋和商业建筑最终是由用户单独委托的。随着项目开发的引入，城市建设进入了一个新阶段，一些机构开始开发更大的建筑项目或整个城市街区。新的发展模式大大加快了城市增长的速度。

城市立法和社会住房的兴起

因为群众住房数量和质量的严重缺乏，在1901年颁布了城市发展法律。这项法律的主要影响是政府资助的社会住房计划的大规模扩张，以及对住宅建筑质量更好的监管。另一个影响是城镇制定具有法律约束力的扩展计划。可以说，这项法律标志着荷兰小镇从有机规划到系统规划的过渡。城市自下而上的发展开始逆转成自上而下的方法。自上而下的规划与现代主义分隔工作、生活、休闲和交通的方式，一同定义了20世纪荷兰小镇的发展模式。

阿姆斯特丹从1200年到2000年的增长：
80%的城市地区出现于1900年后
Growth of Amsterdam from 1200 to 2000:
80 % of urban area after 1900

1938年荷兰公路规划：
国家发展蓝图
Dutch highway planning 1938:
Blueprint for development of the country

战后计划：70%的荷兰建筑物建于1945年以后
Post war planning: 70% of dutch buildings is built after 1945

工业化、现代化和近郊化
INDUSTRIALISATION MODERNIZATION AND SUBURBANISATION

扩张

第二次世界大战（1940~1945年）之前的时期是由危机经济学和历史悠久的城市中心的缓慢增长界定的。二战后的时期成为全球文明史上最长的稳定期和经济增长期。超过2/3的荷兰建筑是在二战后建成的，荷兰人口从1940年的900万增加到2016年的1700万，荷兰的城市地区增加了四倍，这种扩张可以在荷兰小镇的发展地图上清楚的看出来。从阿姆斯特丹的历史边界，就可以看出1900年后的扩张。

汽车和郊区化

在战后规划中，最有影响力的因素可能是汽车拥有量的爆炸性增长，以及完善高速公路系统的引入。直到第二次世界大战，城市建设总是具有一定的紧凑性，并与邻近的城市中心和商业、服务领域存在一定的联系。当汽车标志着郊区化的开始，生活和工作的空间分离，这种紧凑的城市建设情况结束了。在荷兰城市地区的大规模扩张期间，贸易商业和生产移出了城市中心，集中到城市外围的工业区和工作区。值得注意的是，随着建造范围的增加，城市密度明显降低。大部分住宅增长是一种低密度的发展，包括带花园的单个住宅。

历史中心的衰落和城市更新

郊区化和经济活动远离市中心，导致了城市中心的大量减少。在70年代，被遗弃、被忽视的社区，街道和建筑物的问题变得非常紧迫。荷兰政府开始启动计划，以加强和重新定义老旧中心和邻里。这项计划被称为"城市更新"，旨在保持旧区的吸引力，避免原住居民向郊区和新城转移。这个社会经济方案在20年间都行之有效，直到城市再次变为生活和工作的地方。

都市生活方式新认知和再流行

过去20年，人们对城市生活方式，城市中心的生活和工作，尤其是年轻专业人士和富裕公民的生活和工作，有了新的巨大认识。主要原因是服务业的兴起和制造业向国外的外包，这就要求存在不同的城市环境。这种重新认识城市的过程得到了国家政府的大力支持，以及城市基础设施，公共空间改进和密集化战略投资计划的支持。下层住宅高档化和房地产价格上涨的过程，证明旧城区质量背后的强大拉力。在许多城镇中，能够满足不同功能和活动需求的建筑在质量和数量上都大大增加。

如今的阿姆斯特丹：经典城市市中心更受欢迎
Amsterdam today:
Classic Urban downtowns are more popular then ever

荷兰乡村：休闲变得越来越重要
Dutch countryside:
Leisure is becomming more and more important

阿姆斯特丹南阿克西斯区：
21世纪新阿姆斯特丹市中心
Amsterdam Zuidas:
A new Amsterdam downtown for the 21st century

城市边缘鹿特丹：
城市交通，开放景观和农业的融合
City edge rotterdam:
Intermingeling of city, open landscape and agriculture

工业化、现代化和近郊化
INDUSTRIALISATION MODERNIZATION AND SUBURBANISATION

城市升级项目

以升级和重建为主题的政策方法，巩固了重新认识城市的过程。主要应用的项目类型有：

1. 火车和公共交通枢纽。
2. 历史悠久的市中心公共空间和设施的升级。
3. 市中心附近工业区的振兴和改造。

今天的许多荷兰小镇都有精心设计的公共交通枢纽、提升的滨水空间、创意的工厂和21世纪融入生活与工作的办公区域。目前，优质的城市项目案例有阿姆斯特丹kop van Zuid滨水开发和Zuidas开发。这样的项目旨在扩大联系，提供全球商业中心，创造兼具生活和工作的多功能、高密度新区域。

"多城市单一景观"到"单城市多样景观"的逆转

毋庸置疑，荷兰的城市得到了发展，城市间的景观也发生了巨大变化。开放景观和农业区域已经减少，功能也发生了相应的变化。越来越多的乡村和开放景观成为休闲、娱乐和生态区域。最近，很多关于荷兰规划的批评家和思想家统一认识到，荷兰，特别是兰斯塔德地区，发生了转变。荷兰人不再生活在城市景观中，而是生活在拥有很多开放景观的一大片城市地区。荷兰的核心区域现在被看做小镇和城市的连续组合，其中开放景观实际上已经变成了公园结构。这种新现象下产生了一个新术语：城市化景观，指的是城市和乡村的传统边界再也不适用了。

兼具休闲和农业的开放景观

城市扩展和农业开放景观的混合，带来了绿地与周围居民的融合，并且越来越多的转变为休闲经济的绿色环境。过去许多靠近大城镇的农村，已经变成了僻静的郊区，有宽阔的房子、别墅和休闲设施，如餐厅、咖啡店和其他形式的娱乐设施等。

解读都市农业风景

对于荷兰的乡村和风景，有一个理想化的视图，那就是人工圩田区，其几何形状和自然之美振奋人心。这使得人们很容易忽视，大部分荷兰景观服务于高效经济和农业的目标。现在荷兰是世界上第二大农业出口国。农业业务已经大量机械化，并达到工业规模。绿色开放空间不再有经济生产功能，而是转变为景观公园。不仅城市和小镇发生了巨大变化，景观也发生了巨大变化。

成熟的城市和特色小镇

正如前言所述,本书的目的是区分成熟城市和特色小镇,作为城市发展的补充。成熟的城镇是大城市周围的较大都市群,提供完备的工作、生活和娱乐。特色小镇是较小的都市核心,提供便利的生活、教育和休闲设施,有自身的重要特色,而非全面的经济活动。

全球化经济中的差距

毫无疑问,成熟城市的空前普及在荷兰是主要趋势。同样,小镇在偏远省份也是如此。2008年经济衰退之后,荷兰的平均房价下跌了近20%,成熟城市已然恢复。然而,各省的小镇仍在努力寻找出路。一些远离成熟城市的地区甚至面临人口减少和经济衰败。在21世纪的经济全球化中,大城市和小城镇之间的财富和机会差距越来越大。虽然这种差距在荷兰并非不可逾越,但它确实给管理大城市的可持续增长和保持小城镇在经济及社会方面的活力和发展,带来了双重挑战。

城市热点

荷兰城市开发中很好的案例有阿姆斯特丹、乌得勒支、海牙。这些城市有历史的痕迹,同时也无缝融入了现代生活。那里有风格丰富、设施完善的住宅和建筑,以及多样的交通系统。通常,这些城市里有一所或多所大学,一家或多家医院和各种各样的购物环境。越来越多地公司搬到那里,希望更接近他们的伙伴、对手、机场、高技能的城市劳动力和所有好的设施。由于非常便利、机会众多,个人更喜欢高密度社区的都市生活方式。在城市更容易找到工作、挖掘客户,也更容易维护新老朋友的关系。如今,像有孩子的家庭和富裕的老年夫妇这样意想不到的群体,也被城市生活方式所吸引。

成熟城市的挑战

成熟城市的主要挑战是确保它们不会成为成功的牺牲品。通常类似阿姆斯特丹这样的成熟城市,城市管理中存在这样一些核心问题:

1. 作为一个越来越受欢迎的旅游目的地,市中心很大一部分区域正在遭受越来越多旅游带来的困扰。这让当地居民感到城市变得不宜居。大量的游客将城市变成了一个休闲公园,威胁到现有设施、便利交通及合理价格等要素之间的平衡。

2. 城市生活的普及将房价拉伸到前所未有的水平。住房负担导致了阶级鸿沟,并驱离许多人。这对于城市劳动力离开城市、寻找更好的机会影响重大。

3. 为了适应城市的可持续发展,靠近服务区域的有混合功能的高密度社区区域及项目变得至关重要。城市基础设施需要更新和扩展,以适应未来的增长。

4. 城市鸿沟导致居民对日常环境的不满。因此,市政府必须利用市民平台进行互动交流,讨论并适当告知规划和发展过程。

5. 解决当地街道和街区的经济和技术可持续性问题。例如独立建造的市内住宅和绿色屋顶,地下水存储和能源生产。

当代荷兰城市的挑战
CHALLENGES FOR THE CONTEMPORARY DUTCH TOWNS

重要城镇

在荷兰较大城市附近、偏远地区和省份,有许多中小型城镇。通常可以看到,小镇的活力与其在城市网络和经济集群中的位置,以及强大的地方资产,如有吸引力的生活环境、当地遗产和景观质量等,有着密切的关系。在经济全球化的今天,网络之外的小镇变得更加脆弱,更难保持吸引力,更难适应社会和经济的发展。受过良好教育的人都转移到较大的城市,将年长的和流动性较低的人口留在较小的城镇。

特色小镇的挑战

为了保持(特色)小镇的生机和活力,可以设置一套干预措施和政策。像往常一样,这不仅仅是一个城市规划和设计的问题,而是更多地关系到社会和经济战略的问题。

1. 特色小镇的主要挑战是保持其强大的功能,如教育机构的产业集群。同时,特色小镇需要寻找新的机会。与成熟城市不同的是,经济和文化的新趋势不会自动成型。

2. 小镇的主要挑战之一是通过汽车和公共交通保持与区域网络的连接。如今,公共交通或拥挤高速公路的改造进程可能对小镇的活力构成威胁。

3. 当地小镇需要保护好现存设施和服务。年轻人和富人离开的风险,将使小镇设施面临瘫痪。小镇需要设定一个最低标准,保护和利用好那些设施。

4. 对于偏远地区人口负增长的城镇来说,为城市缩小和人口减少等现实问题做好准备。特别是荷兰的边缘地区,如泽兰,林堡和格罗宁根这样的省份,正在制定积极的应对政策。

5. 保护小镇特色是保证小镇独特性的必要条件。荷兰很多沿海的小镇就是因为建设工程破坏了自然美和宜居性,而失去了其原有的价值。

特色小镇分类　FEATURE TOWN TYPOLOGIES

历史小镇　HISTORY TOWNS

费勒　　　VEERE
福伦丹　　VOLENADM
羊角村　　GIETHOORN

水小镇　WATER TOWNS

维恩科文　　VINKEVEEN
多德雷赫特　DORDRECHT
阿尔梅勒　　ALMERE

艺术小镇　ARTTOWNS

贝亨　　BERGEN
拉伦　　LAREN

产业小镇　PRODUCT TOWNS

利瑟　　　　LISSE
阿尔克马尔　ALKMAAR
代尔夫特　　DELFT
恩斯赫德　　ENSCHEDE

科技小镇　SCIENCE TOWNS

莱顿　　　LEIDEN
瓦赫宁根　WAGENINGEN
格罗宁根　GRONINGEN

小镇分类

　　为了更好的理解荷兰城市网络中，与大型成熟城市互补的特色小镇，本书概述了不同规模和不同背景的荷兰典型特色小镇。要深入了解一个小镇，首先对其进行归类。这比小镇的规模和位置更重要。

小镇多样性

　　为了掌握荷兰特色小镇的多样性和起源，我们从历史、经济、文化、基础设施和景观等方面，对其进行归类。这种分类本身并不是科学的，只是为了便于理解。

1. 历史小镇：以历史文化旅游为特色产业的小镇

2. 水小镇：以水系统为特色的小镇

3. 艺术小镇：以文化和艺术贡献为特色的小镇

4. 产品小镇：历史上以当地传统产品为特色的小镇

5. 科技小镇：以科技社区为特色的小镇

拉伦
Laren

羊角村
Giethoorn

多德雷赫特
Dordrecht

代尔夫特
Delft

莱顿
Leiden

特色小镇分类　FEATURE TOWN TYPOLOGIES

主观定位

这种分类方法也融合了公众对小镇的了解。例如：代尔夫特以代尔夫特蓝瓷而出名，尽管如今因代尔夫特大学而被人熟知。正如案例研究，小镇的特征和地位可以随着时间而改变。一些荷兰大学小镇最初是贸易和国防中心，其他小镇是农业贸易中心，但这些小镇后期发展成为艺术文化和休闲中心。本书描绘了这些小镇如今的风貌。很多小镇可以同时被定义为水镇、产品小镇和历史小镇。这种定义方式主要基于读者。然而，为了建立对荷兰小镇的正确理解，我们认为这种分类方法是非常有用的。

历史小镇

是指在发展过程中受阻，并在19～20世纪现代化得到保护的一类小镇。这类小镇通常比较偏远，并依赖于某些特定的资产或特征，如那些随时间消失的贸易、生产或防御。如今，这些小镇通过保持祖辈时期形成的小镇形状、布局和建筑物，保持历史的痕迹。这类小镇也成为旅游景点，带动了当地商业和经济的发展。

水小镇

海洋、湖泊、河流、水道和水系与所有的荷兰城市息息相关。有些小镇与水有典型的关系，主要体现在贸易、国防、交通、水管理和社区发展等方面。这种关系深深的影响着城市设计的结构、公共空间以及单体建筑临水的方向。在许多荷兰小镇，历史形成的土地和水系关系，产生了特殊类型的社区以及关于休闲和价值创造的新视角。

艺术小镇

是指专注文化生产和可替代生活方式的小镇。这些小镇通常在较大的、更具活力的小镇周围，并且有以小镇命名的人和运动。小镇因当地的魅力建筑，景观如森林、沿海沙丘，和便利基础设施，而吸引特定的人群。

产品小镇

当地的产品，如啤酒、奶酪、有技术含量的手工艺品瓷器等，经常闻名先于小镇的名字和名声。小镇名成为了特定产品的品牌名称。这种产品生产的物理环境，如建筑物、贸易中心、基础设施或市场，通常让小镇独树一帜，成为热门的旅游目的地。虽然新产品和服务可能已经超越了旧产品和服务，其形象永存。

科技小镇

21世纪，创新和知识已成为全球经济的核心。有学术研究表明，小镇正成为经济转型的中心，同时有着悠久的历史根基。科技小镇的建设，是为了容纳学生、研究人员和研究机构；同时，创造、提供科技以外的、有吸引力的、舒适的生活环境。

历史小镇
HISTORY TOWNS

来自荷兰的案例
Cases from the Netherlands

18/费 勒　VEERE

28/福伦丹　VOLENADM

36/羊角村　GIETHOORN

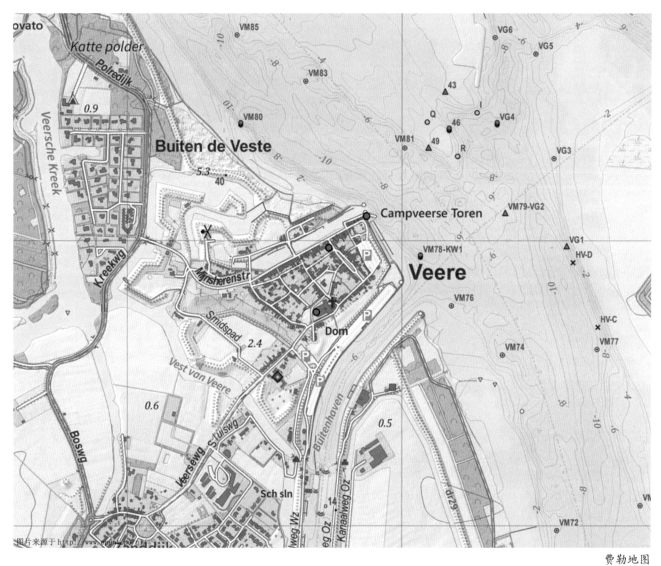

费勒地图
Map of Veere

费勒 VEERE　　HISTORY TOWNS

费勒码头
The Kaai Veere

海天泽国边飞舞的凤凰——费勒

初到费勒，给人的感觉就像《驱魔少年》中那个铃音风之子的家乡。虽然他在书中只是个微不足道的配角并且死去了，但他所在的城镇却让人心神向往。Zeeland（泽兰省）大家很熟悉，是新西兰名字的起源，那位船员把那块新发现的大陆称为NewZealand（新西兰），就是因为怀念自己的家乡Zeeland。不得不说，在心中，故乡永远都是最美好的。就如那位死去的风之子，他常说他要离开这个城镇，要离开单调又无聊的日子，但他最后却把圣洁做成了那颗家门口的铃铛。这样他在异国他乡听到圣洁的铃音时就会想起每次开门，那门口的铃声。叮叮……我回来了，叮叮……有人来了。叮叮……这到底是不是最后的归途……，只是最后的最后，他也许明白了，出去是因为爱，也只有爱才能让他回归。不管在何地，当游子们看到了类似家乡的景色，他虽然把故乡的名字赋予了这个地方，但它毕竟不是故乡。

只是费勒却不同，它让人觉得，初见即是故乡。它与我们的故乡其实并不相像，却让人忍不住想要去触碰，也许触摸的不止是那摇曳在眼前的如画风景，更是那斑驳在每一个角落的历史印记。

1652年费勒地图： 费勒国际海港
Map of Veere 1652:
Veere as international sea Harbour

现成为水上运动中心的费尔瑟湖
Veere today:
Heart of a watersports area called Veerse Meer

费勒小镇街景
The streets of Veere

费勒小镇风景
The landscape of Veere

费勒 VEERE

HISTORY TOWNS

今时今日，沿着那气势如虹的荷兰三角洲工程一路骑行，你对荷兰的印象也许不再只是所谓的奶牛、风车、足球，你会感叹于一路上的如在画中游，也会被荷兰人自强不息的精神深深折服。而费勒，就是镶嵌在费尔瑟(Veerse Meer)湖畔的那一粒明珠。你会发现，历史在这里轻轻诉说，如海风那般随行。在费勒，你会觉得历史从未走远，依然印刻在你的四周。

这个小镇的故事可以追溯到700年前，而今它依然保存完好，这是很让人动容的。如果让笔者来总结，费勒似是一座凤凰般的城镇。因为它在岁月的变迁中，不断地涅槃着。哪怕是此时此刻，它也正在孕育着自己的新生。人们也许很难想象，在这几百年的时间中，费勒已经经历了四次变革，而且这个小镇从来没有退缩和畏惧，面对历史的洪流，它随着时间的浪潮不断前进着。

费勒曾经是一个富裕的国际港口，如今已成为费尔瑟湖周围一个大的水上游乐区。费尔瑟湖曾是一个人造咸水湖，长22公里，内含13个小岛，其中包括六个游艇码头，休闲船达到5000只。费尔瑟湖如今已经是举世闻名的水上休闲旅游之地。水上旅游也已经是这个地区的主体经济来源，这得益于泽兰省曼妙的田园风光和北海海岸线附近那多如珠玑的美丽小镇。

费勒在历史的记载中，最早的定位是防御小镇。所以当人们走在费勒的街头，定会惊叹于这个小镇保存完好的历史遗迹。你可以穿过古老的国防工程的大门，来到那个依然存在的直线型港口，这就是百年码头。而与码头平行的另一边，则是现代游艇的港湾，这里常年停靠着超过100艘的豪华游艇。站在其中，似是可以听到历史回响与时尚潮音碰撞的声音。

码头的街道直接延伸至小镇内部，距离现今公共水域最近的小街道则会一直延伸和扩大至小镇的中央腹地，此处被当地人称为市场，并被诸多的历史建筑包围着。这里承载着费勒几百年的历史沧桑。你也可以去看一看曾经小镇的防御工事。它们都在综合网格中，内有150多个国家纪念碑。这些斑驳的古迹让你与历史似乎只在咫尺之间。

翻开历史的绘卷，13~14世纪时期，费勒只是一个小渔村而已，但却由于其独特的地理位置及环境，慢慢发展成为一个过境海港。1541年费勒成为荷兰和苏格兰贸易往来的官方联系城市。苏格兰人带来了苏格兰羊毛及农产品，用以换取荷兰的屋瓦、板材、绿植和武器装备。随着两地贸易往来的加深，费勒逐渐成为了一个富庶繁荣的小镇。

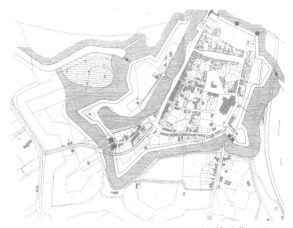

小镇防御工程：
防御工程已经成为大地艺术的一部分
*Fortification structure:
The defense works have become an object of landart*

费勒市政厅
Veere Town hall

费勒市场：过去的市场如今迎合游客的需求
*Markt Veere:
Market place of the past now caters to tourists*

费勒大教堂：依然是城镇最大的建筑
*Cathedral Veere:
Still the largest building in town*

费勒码头排屋
Townhouse Veere

费勒 VEERE

此外，费勒扮演的另一个重要角色是荷兰共和国南部的军事要冲。在与西班牙帝国从1568年到1648年长达80年的漫长战争中，费勒成了荷兰共和国联盟中的重要据点。当时3000名居民（约现在的两倍）中，就有300人是苏格兰商人和行政管理人员。但在1800年左右，苏格兰人为费勒带来的繁荣在法国的统治下很快便灰飞烟灭。费勒从一个贸易和军事重镇，再次沦为一个破败的小渔村，其内港成了只有三十艘渔船停靠的"贫瘠的港湾"。

人们今天所见的小镇主体，大约在1600年左右形成。所有的重要建筑和小镇防御工程都可以追溯到那个历史时期。围绕小镇的防御要塞和港口其实具有双重防御功能，不但可以使费勒免受海洋的侵蚀，也可以抵御外敌的袭击，所谓的天灾人祸俱防。

镇中防御工程大致分为三个阶段。防洪基础墙体是在一系列严重洪灾后兴建的，大约在1500年左右完工。防御工程的第二阶段是强化建设更多的军事系统，因为费勒那时已成为一个重要的据点，提高城镇的安全性成为当权者的重要诉求。第三阶段则是在19世纪初拿破仑一世时期完成的。当时拿破仑决定将费勒的防御工程进行扩展，形成所谓的第二道防线。

而如今，这些往日的军事防御工程却印证着另一种辉煌，甚至可以被看作是人类创作的伟大艺术。它们所呈现的费勒不再只是休闲旅游之地的泛泛印象，而是更为鲜明和硬朗的历史形象。

苏格兰人遗留在费勒码头的排屋，已经成为见证苏格兰与荷兰贸易关系的遗迹。这些房子可以追溯到1540年，当时由苏格兰贸易界的高层发起。现在这些房子成为了费勒博物馆。

费勒市政厅默默地见证了16世纪荷兰人的英勇和智慧。市政厅始建于1474年费勒商业和行政权力的巅峰时期。该建筑采用哥特式风格建造。建筑主外立面以完美的几何顺序，排列着15世纪费勒最重要的7个统治者及其家庭成员的雕像，其中最精致的一个建筑元素为钟琴。历经几百年，却仍然可以运转。旧市政厅如今已经成为举办特色活动的专有设施，尤其作为婚礼场地，深受欢迎。同时，它还展示着费勒许多的历史文物。

费勒大教堂于1479年费勒鼎盛时期开始建设，大教堂的外观、功能和用途在不同的历史时期，都有所扩展和变化。教堂融合了不同的建筑风格，就像费勒小镇本身并不短暂的历史。大教堂最初主要服务于三个宗教团体，每个团体都有一个分区。在西班牙战争期间，教堂丢失了大部分宝藏。19世纪，教堂主要作为医院，部分被拆毁，被作为施工材料售卖。整修后的大楼，现在是费勒的音乐会场和文化活动中心。

1953荷兰洪灾
1953 Dutch flooding

三角洲工程
The delta project

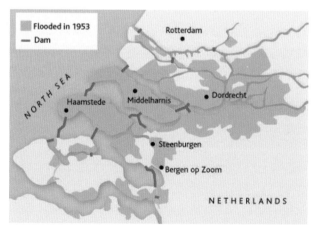

红线是三角洲工程的主要部分
The red line is a major part of the delta project

费勒 VEERE

到此为止，大家也许已经觉得费勒作为一个小镇，经历是何其丰富了。然而，最深刻的变化其实是从近代开始。自19世纪末以来，荷兰对水资源的保护显得不足，对海洋和来自欧洲腹地、通过荷兰河口的水源的管理更欠妥善。一战二战以及20世纪30年代的经济危机，使得水利工程更是长期被忽略，费勒和泽兰省防洪体系的诸多弱点逐渐暴露，水灾风险越积越大。

终于，1953年2月，百年一遇的超级洪水使堤坝一个接一个被摧毁，1000平方公里的荷兰土地被淹没。洪水吞噬了地面的一切。所有的正常生活都被迫中止了。因为当时所有的通讯方式都被洪水破坏了，荷兰政府花了很多时间才了解到这次灾害的严重程度。这次被载入人类灾难史册的洪水夺走了荷兰南部约1500人的生命，也对荷兰海岸线和水资源管理计划的未来发展产生了巨大的影响。

直到1953年大潮灾，费勒仍然直接与北海相连。自古以来，泽兰群岛的海岸长达几百公里，许多小城镇靠海为生。然而这一次灾难彻底改变了荷兰政府的固有认识。通过评估大潮灾害和制定预防策略，荷兰政府开始构想一个惊世骇俗的计划。荷兰固有的地理位置是无法改变，而大自然的气候和洪水也不能预测。就算获得百年安宁，那后一个百年呢？荷兰的子子孙孙呢？

于是一张蓝图在荷兰人手中慢慢绘制出来，并耗费几十年的岁月慢慢铸成宏伟的工程，这就是堪称现代世界十大奇迹之一的三角洲工程。工程有效地调节了从河流流至大海的水源，保护土地不受洪水影响。这一战略对泽兰省的影响是巨大的。许多小镇和村庄因此失去了与海洋的直接联系，像费勒这样的城镇失去了海港以及公海的固有优势。这自然也导致了当地渔业业务的不停下滑，而此工程对环境产生的影响也是巨大的。原来作为航道所用的咸水海域，如今已经变成了风光旖旎的费尔瑟淡水湖。

泽兰省人民并没有因此而消沉，相反，他们迅速应对这所谓的改变，再一次紧跟上了时代的步伐。泽兰省渐渐借由三角洲工程带来的改变，发展成了著名的旅游休闲度假区。在与所有大坝连接的沿岸，休闲旅游度假村如雨后春笋般的涌现。由于大坝的原因，这些度假村的交通也变得十分便利。这使得泽兰省不仅吸引了本国的游客，也辐射向周边的欧洲国家。在过去的30年里，费勒已经找到了全新的经济增长来源，并成为了费尔瑟湖6个休闲港湾的旅游集散中心之一。

提供多项水上运动的费尔瑟湖
Veerse meer offers a variety of watersports

费尔瑟湖区
Veerse meer area

夏季休闲
Summertime leisure

费勒 VEERE　　HISTORY TOWNS

费勒已成为荷兰水上休闲旅游的翘楚，也成为泽兰省附近的德国人喜于选择的度假胜地。许多从周边度假村和度假胜地，像弗劳文波尔德(Vrouwenpolder)，坎珀兰(Kamperland)和鲁伊特帕萨特(Ruiterpsaat)来的游客，也都会选择到费勒一游，哪怕是喝杯咖啡或吃顿小餐。他们有的乘船，有的骑自行车，有的自驾游。但不管如何，费勒都是他们旅行地图上不容错过的一颗明珠。

对于规划设计者来说，非常值得注意的是：这个大型娱乐湖泊与历史旅游目的地结合，不仅仅是一个设计产品那么简单。费勒所散发出的历史余韵，使得这个小镇的整体氛围更像是一种历史的回顾和传递。作为设计者要综合考虑这些文化元素，力求更深入、更令人满意地将设计与历史融合，而不只是简单地完成设计。那些宏伟的防御工事、小巧的港口码头、风景如画的房子、通往大教堂和市政厅的小街道和广场，才是这个小镇的灵魂所在，它们默默见证了费勒的历史，乃至是荷兰的历史。费勒是一个富有历史弹性的城镇，虽然在历史的洪流中，历经四次变革，从贸易和国防重镇，到历经战争，再重新回到一个小渔村，最后变成如今的水上旅游休闲中心。祸兮福所倚，每一次的浴火都让这只凤凰获得重生。费勒正是一座很能展现荷兰人与生俱来韧性的小镇。

编者感言

费勒交的答卷何其生动，不得让人感叹，要想看清历史性的命运，必须看清事态发展的趋势。海天泽国中的三角洲，并不拘泥于过去，它不断地凤凰涅槃着，追随时代和当下，做出了它的选择，并将之尽善尽美。

地理位置、特有资源、历史事件都会影响小镇的发展脉络和命运。城镇历经战火、岁月、繁荣、衰落，却依然鲜活并开出了新的花朵。

在现今的小镇开发和设计中，规划设计师们经常推翻重建，导致许多古迹一去不复返。另一方面，许多小镇又不得不仿古，最终形成了"拆了再仿古"的怪圈。

其实人们应该时刻提醒自己，历史和现代的交相辉映，是这个城镇古今交融的美好尝试。就如费勒这般，不需破坏旧的，却也可以赋予新生。用心与历史对话的规划设计师，其实需要提炼出根植于城镇的文化核心。千百年之前存在的文化图腾，也许现在仍然弥漫在这个城镇中。扪心自问，现在的设计是否能够撑过百年？又或是如何将它所要传达的精神真正地传递出去。

让规划设计如风随行，潜入这座城市，根植于它的脉络中。说起来很简单，其实却是规划设计师呕心沥血的一场苦修。深刻地挖掘地方文化和核心元素，往往只是第一步。如何跟上时代的洪潮，让传承和创新俱在，才能描绘出更广阔的蓝图。

荷兰诗人埃里克·林德纳说："在道路和窗口之间的花园里，我梦想站在那里"。而我说："在历史和现代之间的花园里，费勒就站在那里"。也许我们的花园，就在这里，不光是设计师，也是触碰着这些建筑的人们。

福伦丹地图
Map of Volendam

福伦丹 VOLENDAM　　　　　　　　HISTORY TOWNS

福伦丹海港
The harbor of Volendam

蒹葭苍苍对渔火，荷服飘飘水云间——福伦丹

　　封闭与联通，一念之间。联通是不是就一定能带来更多的机遇，封闭是不是一定要被扣上过时的枷锁？

　　两条大坝，成就了不同时期的福伦丹。这大坝的堤岸，也如人们的心墙一样，封闭、自成一体、难以打开。但人们心中的城镇，不也正如福伦丹一般，它是那么的纯真、繁茂、美丽、生机勃勃。人们自由地在心间描绘自己的理想乡，拼凑着梦想的碎片。

　　如果用唐诗宋词来表述，不就是"晴川历历汉阳树，芳草萋萋鹦鹉洲"①和"谁知闲凭阑干处，芳草斜晖，水远烟微，一点沧洲白鹭飞"②。

　　所以，从福伦丹这个小镇的画卷中，可以看到中西的相通之处。大坝，让福伦丹保留了传统，正如人们心间的那一份青葱。也许随着岁月的变迁，世易时移，心间枝叶繁茂。人们要拨开很多繁杂才能看到最初的那一抹青葱。但是，不论中西，人们都找到了一个对应它的名字。

　　福伦丹配得上这个称号，它是心中的芳草洲。不论是封闭还是围住的岸芷汀兰，芳草洲总是在水的一边。道阻且长，却也，所谓伊人，在水一方。

①唐·崔颢《黄鹤楼》　②宋·欧阳修《采桑子·何人解赏西湖好》

福伦丹和古城Edam
Volendam (on the coast) and Edam (at the back)

1906年的福伦丹
Het dorp Volendam, 1906

福伦丹海港
The harbor of Volendam

福伦丹 VOLENDAM

福伦丹(Volendam)坐落于是北荷兰省(Noord-Holland)，占地面积16平方公里，人口约2.2万。这个昔日以捕鱼为主业的小渔村，也同样因拦海大坝的建成而不得不放弃一直以来传承下来的捕鱼业。但令人欣慰的是，它至今仍保留着昔日风貌。作为著名的旅游圣地，小镇目前每年都会接待几百万来自世界各地的游客。

福伦丹其实是由古城伊丹(Edam)①的发展而建立起来的。1357年获得了城市执政权利的伊丹，是荷兰著名的古城和港口城市。伊丹当时为扩大城市，并让其往港口方向发展，于是决定在东北方向3公里的岛地重新开辟一个新的港口。此外，为了让新开辟之地与南海隔开，政府还专门修建了一条大坝将原来的老港口、当时的伊丹古城和南海分隔开来。这条封闭的大坝就是福伦丹(Volendam)名称的由来②。大坝建成后，渔民们为了更方便地出南海打渔，便聚集于此，有的更搬迁至此。于是，一个著名渔村悄然诞生了。

经过几个世纪的发展，福伦丹的捕鱼业蒸蒸日上，城镇也因此而不断壮大。到了1800年，福伦丹小镇大约已经有大约2000艘渔船常驻，例行一次出海捕鱼的船队就可达到几百艘。当时的渔业对整个荷兰的经济都起到了十分重要的推动作用，后来由于北海运河的开放，南海的渔船也逐渐开始去北海捕鱼。

福伦丹虽然离伊丹很近，但当时整个福伦丹却是被水包围着，只能由水路到达。这种被隔离的居住方式导致了福伦丹居民一直保持着自己独特的生活习惯，不愿被外面的世界打扰，他们几乎不跟外面的世界接触。直到1860年，才开始有外来人进村。不过，也正是这种物理上的隔离式生存方式以及福伦丹人们以己为荣、努力成为"佼佼者"的天性，使他们完好地保存了自己独特的生活习性、自己的方言、自己的民俗和自己的服装，并将这些文化一直延续到了今天。

福伦丹的男人们在艰辛的出海捕鱼路上常常要互相帮助。这种面对大自然时的彼此坦诚和真实造就了福伦丹人友好、爽朗和热情的性格。而福伦丹的女人们则会一边做着传统的民族服装一边等待着男人们出海归来。当期待已久的捕鱼船队满载而归时，整个村子会热闹起来，人们将活捉的鳗鱼立即加工熏制，全村人载歌载舞地庆祝丰收。如今，福伦丹能够涌现出一个个荷兰著名的名歌歌手，也许就源于这个传统。

随着北海运河的开放以及北海边的艾默伊登(Ijmuiden)渔港的建立，北海渔业逐渐繁荣起来。另一方面，接连不断的洪涝灾害给南海周边地区带来了日益深重的灾难和隐患。政府计划建造拦海大坝封闭南海，一来可以彻底解决南海周边的洪涝灾害问题，二来可以在南海地区围海造田，缓解荷兰土地不足的问题。尽管这个惊人的计划由于遭到了福伦丹渔民的极力反对和抗议而一再搁浅，但1916年的一场特大洪涝灾害最终使政府下定决心拦海造田。1932年拦海大坝建成后，福伦丹渔业受到了重创。很多渔民不愿意离开自己热爱的家乡，逐渐改行转业。

---

①伊丹 Edam：荷兰西部的城镇，位于艾瑟尔湖畔。1357年获准建市，拥有著名的奶酪市场。
②荷兰语中 Volle 有封闭的意思，而 dam 指大坝。

福伦丹小镇依坝而建
Volendam town built in accordance with the dam

福伦丹独特的民族建筑
Volendam's unique national architecture

1906年福伦丹传统少女服饰
Little girls of Volendam in traditional costumes, 1906

福伦丹民族服饰的油画
Volendam national costume painting

福伦丹 VOLENDAM   HISTORY TOWNS

福伦丹依坝而居，以渔为业。小镇的建设和规划都围绕港口展开的。在拦海大坝建造之前福伦丹是个名副其实的渔村。而在大坝建成之后，在传统渔业逐渐消亡的道路上，能够转身成为以旅游业为主的著名渔村，甚至被冠以"世界最美渔村"的美誉，关键还是要归功于其独特的文化魅力。文化是能够成为特色的内因。福伦丹独特的民俗民风、民族建筑、传统服装、居住方式等等所形成的福伦丹独特的文化氛围，吸引了来自世界各地的游客。慕名而来的人们，体验、分享着这个平和朴实的小镇里和谐而传统的生活格调。

19世纪初期开始，福伦丹人民就非常忠于自己的服饰。他们的服装漂亮优雅，制作精细，并随着时间的推移逐步形成了独特的服装风格。直到上世纪60年代，福伦丹的人们日常都身着传统服装。目前闻名于世的福伦丹服饰就是遵循那时留下的着装风格，至今未变。

福伦丹服饰，尤其是女性服饰, 深受各国游客的喜爱。它们和风车，奶酪，木鞋一起出现在了荷兰国家宣传片里，成为了荷兰国家象征。福伦丹服饰能获得如此高的国际知名度,跟19世纪前后许多世界著名画家、文学家、艺术家来此写生和采风有着密切的关系。

1873年法国旅行家和历史学家亨利•哈佛(Henry Havard)在旅游传记里第一次提到福伦丹。后来英国著名画家乔治•克劳森(George Clausen)在油画里展现了福伦丹的特色民族服装。接着摄影师争先恐后地来到福伦丹。各种杂志中不断出现渔港边健壮的福伦丹渔民，身着色彩鲜艳民族服装的福伦丹女性，美如油画的木屋前玩耍的福伦丹孩子们。传播出去的传统元素继而又被无数著名画家追捧，使福伦丹成为当时画家们的向往之地。其中斯班德(Spaander)酒店就是当时画家们最喜欢留宿的地方。如今酒店已经更名为斯班德艺术酒店，常年展览着这些著名画家们的真迹，其中大部分画作是当时画家们用来支付酒店费用的。当然这一大批著名画家的到来又进一步提高了福伦丹渔村的知名度。

说到传统服装，就不得不介绍一下，福伦丹的服饰的意义已经不限于服饰本身。它构成了福伦丹旅游的一道靓丽风景，成为了小镇的特色，也形成了荷兰国家的象征。

历史悠久的斯班德酒店可以追溯至1881年
This historical Spaander hotel dating back to 1881

福伦丹特色民族服装
Traditional costumes of Volendam

福伦丹绿色外墙的传统木屋
Traditional wooden house with green walls

福伦丹照相馆保留着林青霞年轻时曾在此留影
Photos of Brigitte Lin at Volendam Photo Studio

福伦丹 VOLENDAM

女性服饰要从头上的风帽说起，这是一种非常珍贵的蕾丝风帽，将风帽挤压成型也是一门艺术。最显眼的是脖子上三圈带有金锁的红珊瑚珠项链以及胸前美丽刺绣图案的围兜，还有挂在脖子上的蓝白相间的领带，下端是流苏，这是福伦丹女性服饰的经典组成部分。围兜外面穿的是黑色棉质或羊毛半卷袖短上衣，上衣对襟开口用隐蔽的钩圈闭合，再由一根蓝色的羊毛线修整拉平。下面是长裙，色彩比较多样化：红色条纹裙或者七彩裙第一次穿必须在婚礼上，以后只在节日时穿；周日穿灰黑蓝条纹裙；做针线时则穿黑色的皱褶裙；干活时穿纯蓝色裙子。长裙外面是围裙，普通围裙大多是蓝色棉布条纹，节日时穿的围裙用黑色羊毛料。腰部有刺绣图案，和上身围兜的刺绣呼应，显得更加漂亮。

男子的脖子上围着黑色亮十字绣图案的领子。两粒金质领扣表示成年，未成年男孩用银质领扣。衬衣带有红色或白色条纹，无袖，双排扣。外衣为黑色羊毛料，有6粒钮扣，只有下面三粒扣紧。袖口也会用同样的钮扣扣紧，领口处两边都可以挂银质挂链或其他挂件。所有上衣都塞进长裤腰内，长裤一般采用黑色丝绒，裤管很肥，裤腰前是一块正方形前盖，用两粒银质钮扣缝在裤腰上。男子的脖子上带着黑色丝质形如鳗鱼的领围，领围末端打结后塞入外衣领口。这种黑色丝质领围也是福伦丹男子服饰的经典组成部分。

走在福伦丹的大街小巷，看到穿着传统服装的当地人，从规划统一的绿色外墙的传统木屋里出来，用福伦丹方言互相寒暄问候，宁静中透着古朴。人们来到这里，可以体会到福伦丹的所有特色，都不是刻意为之。这种特色有着深厚的历史文化底蕴，作为福伦丹人的骄傲已经深深融入到血液中。

编者感言

当看到福伦丹的故事，我们对这个小镇的定位就早已超出了风景之外。它生于大坝，却也没有因为另一条大坝而消亡。这倒是让我们想起了：围城之内，围城之外。

因为封闭，福伦丹保留了自己独特的文化。而在当今这个全球化的世界中，这些文化却反而显得那么独树一帜、弥足可贵。就是因为渔夫特有的纯朴和敬畏之心，福伦丹并没有那句"围城之内的人想要出去，围城之外的人又想要进来"所展现出来的无奈。

从某种意义上来说，福伦丹甚至成为了许多人心中的理想乡、黄金之乡、应许之地。正如中国古代文人所传唱的那一片芳草洲那般让人向往。

那里，岸芷汀兰，郁郁青青。守着最原始的纯真，守着许多人的最初。这也许就像我们的那一块心地，纵有心墙阻隔，却又让人神往。

现在这样的时代，却依然有这样的小镇，它以最宁静的面貌展现在大家面前。

这个时候真的很想吟问一句：封闭就真的不好么？

有的时候，它是不是也算是一种坚守。

不过，融入却也能带来新的感动。

如今，千千万万的游客造访这座小镇，他们融入到福伦丹的画卷，带给这座小镇新的气息，就如新版《魔女宅急便》的插曲——"回忆的缺口"。能不能有那么一个地方，可以放下我们的回忆，却也留下一个缺口，让我们能随时返还。答案已经不言而喻了。

羊角村地图
Map of Giethoorn

羊角村 GIETHOORN　　　　　HISTORY TOWNS

羊角村的河与桥
Canal and bridge in Giethoorn

一首悠扬绵长的舟上乐歌——羊角村

如果说海是荷兰人迈向世界的征途，那泽就是流淌在心间的那一份独存的诗意。如果你要看荷兰最美的沼泽，那么一定不能错过羊角村。

水盈之处，恩泽润物。羊角村就是承载着这份绵长余韵的地方。为什么亚洲游客特别钟情于羊角村？行舟于水上而吟，这是亚洲许多国家文化中共通的浪漫，如一首欸乃船歌，唱彻古今。

划着小舟，穿梭在小河之上，心中有着忐忑，却被岸边不停更迭的美妙景色所吸引。这样的河流往往如生命一般动人。它的两岸有着看不完的风景，数不清的故事。你永远不知道下一刻你将会看到什么，遇到什么。也许时隔多年，你依然记得那一趟划艇之旅：一个人在划桨的韵律中哼着的为自己壮胆的歌，也记得那一片在河岸便采的野薄荷，将它放入口中，听到了岸边屋子的花园中传来的孩子们的嬉笑问候之声。

行吟泽畔①，随遇而安②。这是《楚辞》和《易经》中的人生哲学和态度。文化与诗意也如这奔流不息的河水一般润泽着我们的心田，直流入心底，从未停歇。

在羊角村，不再是"泽畔有人吟不得，秋波渺渺失离骚"③，而是"泽畔有人吟可得，秋波渺渺题离骚"。

泽雷随，乘马逐鹿，舟行水上，随遇而安。

①屈原《楚辞·渔父》②《易经》泽雷随，随卦　③《鲁迅日记》1932年12月31日《无题》诗

小镇周围的水系
The water around the town

小镇周围的水系
The water around the town

连接村庄的小桥
Small bridges connecting parcels

运河两侧遍布迷人的老房子
The charming old house on both sides of the canal

羊角村 GIETHOORN

一个小镇被用作电影场景后，是否有可能依然如初？当然可以！荷兰电影《吹奏》(fanfare)的拍摄让羊角村名声大噪。而今如若再去羊角村，会发现如今的景色与电影中并无二致。

时至今日，羊角村已成为Weeribben-Wieden国家公园旁边的一处国际旅游胜地。这个小村庄原来是一个偏远的农业社区，由于土地并不适合建设道路，便慢慢形成了遍布村落的发达水系，而后演变成了通过水路连接的小岛，上有农场、房屋和各种设施体系。

在一个国家中，各个地区的现代化进度往往是不均等的。与大都市和以工业化引擎为代表的进步相背离的，往往是带有时间停滞印记的"遗留村落"。这些遗留村落的文化、习俗和仪式得以存续，景观也不曾有多大改变。羊角村周围艾瑟尔河(IJsel)的景观就是这样的一个例子。

羊角村是一个位于大规模农业地区中的小村庄。由于其独特的地理环境，在这里，时间就像静止了一般，不曾来打扰过这个小镇，历史的风貌鲜活地保存了下来。

很久之前，羊角村对于大家来说是一片遥远的森林和泥炭沼泽之地。这个区域的人口密度非常低，主要集中在几个孤立的定居点和村庄里。在早期，泥炭是村庄收入的主要来源之一。当时，泥炭通过小船被装载运输到附近城镇的市场上售卖。由于泥炭的过渡开采导致了景观的严重损失，慢慢变成了只有陆路和水系的有机马赛克纹理地貌。

最早的居民带着这样的开采模式在这里定居。17世纪，羊角村形成了相对固定的形状和位置结构。随着资源的减少，泥炭开采逐渐淡化。在17世纪50年代左右，当地的经济变得更加多样化。随着家禽养殖、甘蔗种植等产业的扩展，农民、渔民和甘蔗收割人员成为主要人口来源。

羊角村的陆地都被水路所分割，陆路交通不可能完成所有作业。这就导致了羊角村的内部交通不得不都交由小船来完成。小船可以运输任何日常用品，从新鲜的面包到村民葬礼的购置等。对外运输则主要包括了农产品、甘蔗、芦苇等相关产品，通过小船可以方便地运到附近城镇的市场。直到20世纪初，该村依然与任何道路系统相断开。即使在今天，羊角村也没有实现汽车通行。

荷兰电影《吹奏》
Movie 'Fanfare', 1958

Weerribben-Wieden国家公园
Weerribben-Wieden National park

羊角村 GIETHOORN

HISTORY TOWNS

20世纪50年代，羊角村因为荷兰历史上最成功的一部电影《吹奏》，而被人们从抽屉的角落中翻找出来。

这部电影的导演伯特·汉斯特若(Bert Haanstra)是荷兰早期电影史上的殿堂级人物。他的作品不仅在戛纳电影节上获得大奖，晚期的一部关于玻璃生产的纪录片更是赢得了奥斯卡奖的青睐。

《吹奏》这部电影主要描绘了两个互相竞争的狂热乐队。他们在当地音乐节上相互角逐，甚至因此让村庄产生了分裂。两个乐队都在羊角村的谷仓里练习，并在船上进行巡回演出。在节日那一天，两个乐队在舞台上的演出同时结束，音乐却奇迹般地融合在了一起，不仅赢得了奖杯，还实现了整个村庄的再次团结。

导演伯特·汉斯特若设法简化荷兰特征和荷兰文化的本质，于是采用了一些辨识度高的人物和独特的村庄背景为承载点，再配上适当的幽默和曲折作为调和剂。

汉斯特若会选用羊角村作为电影的背景，多多少少是因为他本人出生在上艾瑟尔省(Overijsel)。在那之前，羊角村几乎不为人们所熟知。但由于这部电影产生的巨大反响，数以百万计的荷兰人看到了荷兰文化的特殊意义。羊角村一夜之间变成了热门的旅游目的地。旅游也很快成为了村民最重要的收入来源之一。随着当地农业的消失，许多人选择来到这里，购买一处废弃的农场，作为第二居所，用以享受休闲时光。

因为村庄被电影发掘，周围的景观也逐渐成为热点，立为国家公园，并被命名为Weerribben和希特霍伦湖(Giethoornse meer)。自然保留的景观大小约35平方公里，在欧洲同等尺度和质量的自然沼泽风貌景观中，算是独一无二的。保护区内景观视野开阔，水面能折射出绚丽的光影，还有许多珍稀的动植物，是理想的骑行、划船、徒步旅行、体验自然的好去处。Weerribben是很多荷兰人熟悉的度假胜地。得天独厚的沼泽风光已经让Weerribben和周边地区多次被评为"荷兰最美丽的地区"。

上个世纪50～60年代，地方和省级相关部门做了很多关于羊角村的整治开发计划，旨在为当地的农业、工业和交通运输提供更好的发展机会，但都被国家政府否决了。那时，国家风景和文化委员会认为羊角村周边地区的人文、风景太珍贵独特了，不能被残酷地现代化。许多人会认为这是电影的明星效应所致，其实所有这一切都发生在村庄因那部电影成名之前。也正是因为国家政府的抉择，使得羊角村和周边景观都保持了它的历史特性。如今羊角村是著名的历史保护遗迹（1981年），需要保持其独有特性和景观记忆。国家政府的这一决定促成了后来羊角村旅游业、休闲经济多年的蓬勃发展。

连接村庄的小桥
Small bridges connecting parcels

羊角村小船的使用已有历史
Historic use of Giethoorn flat boats

泛舟是游览羊角村的乐趣之一
Pleasure boating in Giethoorn

羊角村出售的新船
New Giethoorn boat for sale

羊角村 GIETHOORN

羊角村的布局保留了历史布局方式，由纵向的三部分区域组成，分别为北区（Noordeinde），中央区（Middenbuurt）和南区（Zuideinde），由河道相连接，通过小船通行。村中典型的茅草屋被称为"Bump农场"。农舍宅屋和农仓谷仓共享一个大屋顶，连成一个整体。每个农舍被水道隔开，通过小桥与外部连接，小桥高度足以让小船通行。

羊角村内有近200座小桥。船行文化作为村庄悠久的生活方式，成为了一种历史遗存。村中有各种类型的船只，满足不同的需求。有用来载人的，也有用来运输小物件和大件物品的各种船只。船只的设计非常平坦精致，使它们能够通过相对低矮的桥梁。在较小的水域，可以划船或撑篙；在较大的水域则可以航行。如今，船只主要靠电力来驱动，可以保持村庄宁静的生态环境。

令人惊奇的是，这个魔力四射的小村庄，特别受亚洲游客的喜爱。如今已有超过20万来自中国的游客，到访过这个有"北方的威尼斯"美称的荷兰水村。在他们的行程表中，羊角村与巴黎、阿姆斯特丹和伦敦这样的大都市一起被列为他们欧洲旅游的目的地。随着时间的推移，旅游业已成为当地的主要经济支柱，小餐厅和游船租赁服务成为主要经营方式。但羊角村依然没有修建过多的道路和基础设施。

编者感言

许多时候，美好的东西往往就在我们的身边，当我们和它们时时刻刻在一起，总会视为理所当然。而当有人将它们拾起，我们才发现，原来这才是最美的特别。

所以，从另一个方面来说，羊角村是幸运的，电影让人们看到了它值得我们去保护和缅怀的一面，也使得小镇更加有效地被保护起来。其实因为影视而带动地方的旅游业，在中国也早已有很多成功的例子。

不过，扪心自问，如果羊角村在中国，是不是也依然能被保护的如此完美呢？我们是否也能见到如此这般的羊角村？其实这是政府和规划师们都需要去面对的问题。

当一个小镇成为热点和争相旅游的目的地时，政府要以怎样的方式把握热度蔓延的节奏，促进发展，是至关重要的。而对于规划师来说，保持原汁原味还是锦上添花，更需要拿捏准确。一味追求速度和过度引爆带来的不仅仅是审美疲劳，也许还会给小镇带来不可逆的永久性改变。

正如篇首语所说的，羊角村就如一首悠扬绵长的舟上乐歌，泽畔有人相和，歌声重叠。一路行来，我所学会的，是易经中的随卦，泽雷随。有的时候，有一些东西，我们要随缘，随遇而安，行至便得。不要拔苗助长，也不要过度强求和热炒。失去了本真不说，还失去了那份最美的特别。

水小镇
WATER TOWNS

来自荷兰的案例
Cases from the Netherlands

46/维恩科文　　VINKEVEEN

54/多德雷赫特　DORDRECHT

64/阿尔梅勒　　ALMERE

维恩科文地图
Map of Ronde Veenen

维恩科文 VINKEVEEN　　　　WATER TOWNS

维恩科文的水景
Waterlandscape Ronde Venen

一张一合如绿肺，一吐一纳似海绵——维恩科文

初看这个区域的地图时，总有一种不敢相信之味。这样辐射状的水陆，我总觉得只有上帝才能造就。所以说，时间可以成就许多不可能为可能，就如愚公移山一般。但还是有所不同，愚公是一心要移开这座山，而此处则是人们为了生计生活，在采挖泥炭时分隔生产区与生活区而慢慢形成的楔形地貌。

虽然有人说维恩科文是一个历史的讽刺，不过我们却看到了它的另外一面。几个世纪以来，正是维恩科文人一直用勤劳和智慧铸就了这样的奇观，而后成为风景名胜。

贫穷和富裕，如此鲜明的对立面，却在这里开出了两生花。改变之间，是时间和历史的轻轻流淌。这不仅让我们想起了一句话，"人生中没有哪件事是徒劳的"。也许现在你觉得自己做的事情微乎其微，但请不要因此而放弃努力。维恩科文给我们呈现了一个最好的例证。

维恩科文地区风景
View on the vinkeveense

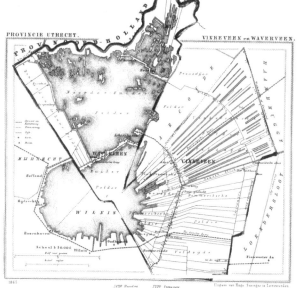

19世纪景观几何分析
19th century analysis of landscape geometry

18世纪圆形景观结构地图
18th century map of circular landscape structure

维恩科文 VINKEVEEN　　WATER TOWNS

维恩科文是阿姆斯特丹以南De Ronde Venen圩区的众多小镇之一。这个城镇的名字源于惊人的景观几何形状。圆形的水系，渗透在绿色的楔形地块和数个小村庄中。该地区曾是阿姆斯特丹地区的泥炭地。那里的泥炭土地曾被开采和销售，留下的是不规则的陆地和水系。这种水陆相间的独特地貌如今成了著名的水上运动湖区和独特的岛屿，也逐步成为了阿姆斯特丹地区的富人别墅区。

荷兰人为了生存，在改造自然的过程中，逐步形成了独特的人造景观。农业生产和治水管理使自然景观慢慢地渗入了人文内涵。阿姆斯特丹以南著名的De Ronde Venen圩区便是这种景观的典型例子。为了农业发展，此处的土地被分割成许多小块，其尺寸主要是根据早期农场的规模和牲畜的数量。为了保持土地的干燥，每个地块都有垂直于弯曲河流和水系的纵向沟渠。因此，在13~14世纪出现了楔形地块的形式，每个地块都有较宽的一部分用以连接河流，还有一小部分则面向土地中间位置。加上古老的曲线形水系，一种独特的景观渐渐地形成了。几个世纪以来，由于农业活动的影响，土地变得越来越低，该区域越来越难以维持水土平衡，也影响经济发展。因此，早期的荷兰规划土建工匠运用他们的智慧将开采泥炭的水道疏通连接，形成了新的运河体，既创建了先进的水循环系统，将多余的水流引入大海，也形成了独特壮观的风景。

景观形成的另一个强大力量则是泥炭开采。1550年至1950年之间，荷兰有了丰富的泥炭产量。这些地区经改造成为了10万多公顷的自然景观。在14~15世纪，De Ronde Venen周边地区的泥炭被作为各种燃料卖给了阿姆斯特丹这样的大城市。将泥炭运输到城镇里是各家各户的一件大买卖，需要大型船只和集市，但更重要的是那些城外的泥炭开采者。对于维恩科文以及De Ronde Venen的当地人而言，泥炭开采是解决了收入问题，无形中也给景观带来了长远影响。

随着泥炭开采，水域面积越来越大。这对土地来说是个危险的信号，并且土地本身随着农业活动已经产生了越来越多的沉降。在这样的情况下，有必要采取措施使剩余的土地保持干燥。因此，当地很早就开始利用风车来进行排水。如今在De Ronde Venen周围仍然可以看到各种残存的风车。可以想象，管理像维恩科文这样的城镇，其中一个重要问题就是如何保持土地干燥，以维持经济生产，满足居民需求。如今，水利局充当起了这个管理者，统一负责水资源管理、饮用水生产和供水安全等，也是荷兰政府管理系统中重要的子系统。

维恩科文临湖的私家别墅
Exclusive villas in Vinkeveen

维恩科文湖风景
View on the vinkeveense

维恩科文 VINKEVEEN　　WATER TOWNS

　　De Ronde Venen的地理区域总体是一个村庄群落的概念。自19世纪90年代起加入行政区划，这里逐渐形成了一个多核城镇。各个村庄在发展过程中逐渐合并成了小镇，这使政府的区域管理更为有效，并降低了成本。小镇的核心主要包括阿布考德(Abcoude)，迈德雷赫特(Meijdrecht)，班布吕赫(Baambrugge)，维尔尼斯(Wilnis)和维恩科文(Vinkeveen)。其中，迈德雷赫特是最大的核心，拥有大量工业和居民定居点。Ronde Venen东部边界是国家高速路（A2线），是荷兰最重要的经济动脉。阿姆斯特丹和乌得勒支高速公路的附近区域成为了各类企业和居民及区域导向的焦点。Ronde Venen是荷兰中部城市网络的重要组成部分。在那里，人们可以获得很多机会，同时还有优美的居住、生活环境，尤其是娱乐环境。

　　De Ronde Venen最著名的区域要属维恩科文。泥炭开采对景观产生了巨大影响，留下许多美丽的湖泊和星罗棋布的小岛。它们在很长一段时间内都未产生经济效益。然而，随着休闲产业的兴起以及荷兰人个人财富的增长，湖泊重新有了用武之地。这里因为各种水上运动，如帆船冲浪、划船和潜水等而大受欢迎，成为了现在人气颇旺的水上运动区。现今，由于周围区域内的许多家庭拥有游船，他们会在假日选择水上运动或徒步旅行等活动。于是小的岛屿变成了私人度假的乐园，到处都是度假屋。19世纪90年代鼎盛时期，许多大型豪华别墅建在较大的岛屿上。由于水上运动和豪华别墅的兴起，维恩科文已成为荷兰一个知名度颇高的名字，也是全国唯一一处融合了水上运动、奢华别墅、迷人风景的地区。许多成功的荷兰人都选择定居在维恩科文。慢慢地，移居维恩科文已是拥有成功职业生涯和巨大财富的代名词。这里最著名的居民应该算是荷兰传奇足球运动员约翰·克鲁伊夫(Johan Cruijf)。他在这里生活了20年，是19世纪70年代左右第一批购买沿湖别墅的居民。

水上休闲运动
Watersports recreation

维恩科文湖的休闲岛
Recreation islands of Vinkeveen Lake

水上休闲运动
Watersports recreation

维恩科文湖上的游艇
The yacht on Vinkeveen Lake

维恩科文 VINKEVEEN   WATER TOWNS

维恩科文湖汇聚了三个湖泊，约10平方公里，拥有荷兰最干净的地表水。湖泊的形状、小岛的位置和周边的条形土地是长年累月由于泥炭挖掘而慢慢形成的。20世纪，有些地区还被用来进行砂石开采，如今部分水深足有53米。纵观全貌，这里的土地和水形成了魔术般的空间结构。这是连设计师都无法想象和刻意创造的。有趣的是过去的景观在新时期也有新的用途。如果坐在船上穿越该地区，可以看到大量的灌木、芦苇、鸟类和鱼类等生物。水域周围也散布着一些休闲码头和露营场所。湖中有许多私人岛屿，从附近码头乘船可以直接到达岛屿的休闲亭。有些岛屿也具有商业和卫生设施。

该地区的主要轴线之一，即湖泊之间的一个特有分区，是连接温克文和巴姆布鲁日的一条狭长小路。这种景观元素被称为"De Baambrugse Zuwe"，因其临湖的别墅地块而闻名。别墅占地在2000～5000平方米之间。建筑风格融合了经典、现代的元素。所有别墅都临水而建，设有游艇停泊的私人码头。虽然有着关于这块地区的体积、高度以及街道形象的规划限制，但总体来说，别墅、岛屿和湖泊的整个空间组织是经过几个世纪有机发展进化而成的产物。另外有个关于维恩科文的玩笑说法："维恩科文是少数快乐人的巢穴，甚至是人们炫耀财富的象征"。不过，毫无疑问的是，这里的景观给所有人都带来了欢乐。

可以说，维恩科文的对比显示了历史的讽刺。在农业活动和泥炭收割期间形成的景观，是数百年来辛勤工作却拿着低工资的人们创造的。而在新的时代，由这些艰苦劳动所产生的相同景观，却已在不经意间成为了富人和名人们寻找休闲时光的风景地。

编者感言

俯瞰维恩科文区域，在脑中始终萦绕的是两个字："楔形"。其实对楔形并不陌生，是因为汉谟拉比法典上的那些古老文字和两河流域的灿烂文明。但是只看文字，的确是无法很好地理解。直到看到维恩科文以及Ronde Venen区域的地图，才终于让人明白这个楔形地块的真正含义。它更像中国的扇面，由中心辐射出去。虽然它的形成是因为特殊的地理环境和经济活动，但还是不由得让人感叹于这鬼斧神工。

无心之功总会比刻意为之好上那么一点点。而这一点点，让人觉得更多是属于一种超越人力的东西。也许是自然，也许是宇宙，也许是神明。但最后，这还是人类所完成的。人的灵性，也许才是造物主最舍不得的一脉。化腐朽为神奇，无心也能从自然中剥离出发芽的种子来。

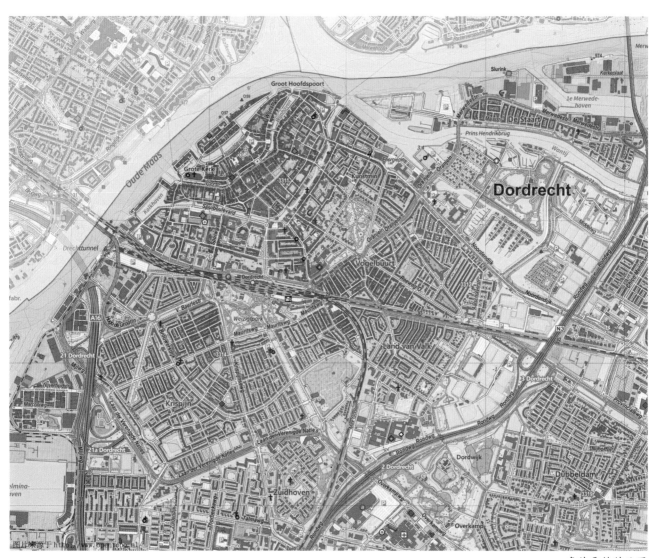

多德雷赫特地图
Map of Dordrecht

多德雷赫特 DORDRECHT　　　　WATER TOWNS

新多德雷赫特港
Nieuwe Haven Dordrecht

西南形胜，三河都汇，水镇自古繁华——多德雷赫特

多德雷赫特很符合人们想象中的几百年前欧洲小镇该有的样子：发达的贸易市场，复古的建筑风格，不同的宗教教堂，各种行政人员穿梭在市政厅，有条不紊地忙碌着。遥远国度的港口，古老工事的海滨，叙说历史的城门，昔日的街巷，教堂的高塔，在水波中摇荡的一弯明月……一切似乎都没有改变，历史扑面而来。

走进一间街口的酒吧，依然是木桶和木色搭配得恰到好处的吧台。人们不禁畅想，下一秒开门而出，是不是就能穿越到那个年代。瞬间，外面已经变成繁华的街市，河上是满载货物的船只，人们在河口交易买卖，称量着金币。经济和贸易在这里生根发芽、开花、结果。

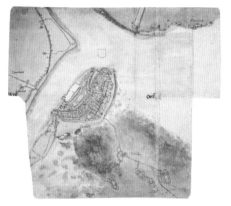

1565年左右的多德雷赫特
Dordrecht around 1565

1619年扬·凡·戈因《通往多德雷赫特的贸易船》
Trading ships on the way to Dordrecht, 1619. Jan van Goyen

1649年多德雷赫特地图：荷兰水城防御工事
*Map of Dordrecht 1649 :
Dutch watertown fortification*

Dordse会议：17世纪多德雷赫特宗教会议
*Dordse Synode :
Religious convention in Dordrecht in 17th century*

多德雷赫特 DORDRECHT

多德雷赫特（Dordrecht）是荷兰最古老的水镇之一，坐落在多德雷赫特岛上，被荷兰三条著名的河流诺德河（Noord），梅尔韦德河（Merwede）和旧马斯河（Maas）围绕。历史上，多德雷赫特作为荷兰南北之间的贸易枢纽发挥了重要作用。当贸易和物流中心逐渐向海边转移到鹿特丹港口时，多德雷赫特也因此失去了它的鼎盛繁华；然而因为地处水边，造船业和运输业仍然很发达。

多德雷赫特坐落在目前欧洲最繁忙的三河交界处。在这里，梅尔韦德河被分割成诺德河和旧马斯河，分别通往鹿特丹和北海。多德雷赫特的故事可说是河流的故事，是河流上行驶的来来往往的船只的故事，也是事关贸易和运输经济的故事。虽然多德雷赫特逐渐失去了荷兰最大港口的地位，但时至今日，人们依然可以从这些属于多德雷赫特400多年前鼎盛时期的精彩故事中，窥见当日之一二，不由得让人肃然起敬。今天，人们很难相信跟鹿特丹交通连接便利的多德雷赫特在其历史发展中只能通过船只进入。事实上，直到1936年多德雷赫特才建成第一座24小时自由进入小镇的长桥。正因为如此，附近流传着的一个经久不衰的逗趣之语："多德雷赫特人思维独特，是由于他们世代生活在岛上"。

很久以前，多德雷赫特就是诸多贸易路线的交叉点，不论是从安特卫普到阿姆斯特丹，还是从布拉邦到鹿特丹，都必须经过这里。那个时候，陆路交通不太可靠，因为冬天会封路；而火车运输还承载不了大批量的物资。与之相比，水路运输自然而然地成为了主要途径。多德雷赫特是一个理想的中转港口，货物卸载后储存在各大仓库里，然后再转去各式各样的市场上交易。当时多德雷赫特的贸易、水路交通、港口物流以及专业劳动力的一体化程度相比荷兰乃至整个欧洲来说都算得上是佼佼者。藉由水上交通优越的地理位置，多德雷赫特发展迅速，变得越来越强大。14世纪，被荷兰政府赠与具有征赋交通税权的"荷兰县"的称号。这种特殊的地位允许它对这里卸载的每一笔交易征税。通过这样的方法，多德雷赫特获得了扩张所需的资本，并投资于当地基础设施、地方政府和地方福利。因此，这里的商人生活富裕，也建造了许多美丽的豪宅。留存至今的许多公共或私人建筑无不在向人们展示着昔日主人的财富和辉煌。

多德雷赫特 Visstraat
Visstraat Dordrecht

多德雷赫特沃维沃斯港
Wolwevershaven Dordrecht

新多德雷赫特港
Nieuwe Haven Dordrecht

多德雷赫特葡萄酒港
Wijnhaven Dordrecht

多德雷赫特 DORDRECHT

小镇的中心面积不大，只有2平方公里左右。尽管如此，人们仍然可以瞥见多德雷赫特中世纪时期如同迷宫一般的街道。街巷、码头和港口的景象依然保存完好。小镇的主要交通结构是两条同心的环形水路，船只能由此进入。第三条环形的水路则是以前留存的防御工事。当时人们试图用城墙包围城镇，所以在城镇内创造了非常有效的运河和港口系统。因为建造城墙耗资巨大，而且土地资源不足，所以小镇必须设计得紧凑而高效。而圆形可以以最小周长创造出最大面积，所以当时的小镇规划呈圆形布置。如果我们翻看旧地图，会发现在梅尔韦德河和旧马斯河的角落，所有建筑物都沿水而建，而其他的基本农业和生产设施则被安设在城墙之外。

不难发现，小镇的功能结构是根据当时交易的产品类型及其需要的空间和设施种类来具体划分。街道、码头和港口的名称清楚地体现了当时各种产品贸易区在小镇内的分布形式。酒类港口在酒市场附近；肉类市场在接受猪肉的港口附近；建筑材料市场则在石灰港口附近。另外，一些特定的产品被安排在小镇外围。比如桐油市场被设置在船只进入小镇的主入口处。由此可见，小镇的功能区域是由产品和行业类型决定的。总体来说，多德雷赫特以及荷兰许多其他的水镇整体上都是组织良好的多元化市场的整合，似乎就为贸易而生一般，整个城镇就是一个庞大丰富的市场。水路交通、卸货码头、仓储以及生活设施的一体化程度极高。

在老地图上，还可以解读到这个经典的水贸易城镇的一些有趣的细节。比如每个城门都有特定的用途，有专门用于运输乳制品的城门，专门用于运输袋装货物的城门，也有专门供地方军队使用的城门，还有专门供牧师使用的城门。由此可见，多德雷赫特有一套严格的管理制度来规范从水路以及城门进入小镇的方式。

此外，在面积不大的多德雷赫特镇有超过10个不同宗教信仰的教堂。不难发现，教堂和宗教在日常生活和社区组织中发挥了极其重要的作用。小镇的中心地区还是许多社会公共设施的所在地，比如老人院，病人服务中心以及地方政府。所有一切都说明了当时多德雷赫特高水准的城镇组织能力，政府机关有一套严谨的行政管理制度以确保小镇高效且安全地运行。多德雷赫特这套高水准的城镇组织能力也同样出现在同时代的其他荷兰城镇中，而这套独特的城镇网络体系主要形成于16～17世纪的荷兰共和国。

多德雷赫特大教堂
Dordrecht Grote Kerk

梅尔韦德河
Merwede

梅尔韦德河上的桥梁
Bridge over the Merwede

多德雷赫特 DORDRECHT　　　　　WATER TOWNS

伴随多德雷赫特城镇发展的也不仅仅是好运,两次大灾难的来临严重影响了它的历史进程。例如众所周知的1421年大洪水,彻底改变了荷兰的版图,多德雷赫特因此与内地切断开来,成为了一个岛上城镇。因此,通过陆路实现的贸易量骤然下降,贸易利润也随之减少。而1457年那场毁灭性的大火让多德雷赫特几乎变成了灰烬,城镇的发展也因此严重受挫。

尽管如此,那个时期的多德雷赫特仍然是荷兰最先进和最强大的城市之一。在后来管辖荷兰地区的城市集会上,多德雷赫特得到了第一把交椅,其优惠条件就是可以为荷兰地区提供一名主要管理者,另一个好处则是让城镇对宗教事务拥有主导权。因此,1619年多德雷赫特成了荷兰历史上最重要宗教会议的举办地。那次会议历时200天,世界主教会议对宗教的地位和诠释有着极其重要的影响。来自荷兰和国外的代表一起商谈圣经的正义诠释,统一了圣经的正确翻译,而这个版本也一直沿用至今。贸易上的得天独厚使多德雷赫特成为荷兰共和国最富有的城镇之一。当然这种优势随着阿姆斯特丹和鹿特丹这样大城市的崛起而慢慢落下帷幕。在黄金世纪,多德雷赫特没能跟上海上贸易的步伐,从而逐渐衰落,到17和18世纪最终变成了一个只有丰富历史回忆的贫穷孤岛小镇。

直到19世纪后期,随着工业化的崛起,多德雷赫特才开始有了一些发展。当蒸汽火车网络开始建设,多德雷赫特在1872年开始与外界连接。1885年在旧马斯河上架起了一座桥梁。火车站的设立意味着给城市发展注入了新的活力,在城外很快发展出了一些新的社区,城市的南边增加了新的工业港口。这些新的基础设施使多德雷赫特成为了梅尔韦德河和诺德河区域造船业、金属工业以及土木工程行业的产业聚集区。如今,海事工业综合体是多德雷赫特最重要的经济支柱。经济活动再次极大地影响了多德雷赫特。伴随着城镇的发展,城镇的道路以及中心历史核心区域向人们揭示了不同时代的建筑风格以及经济活动类型。随着城市的扩大,多德雷赫特也逐渐从经典城市转型为现代城市,居民区慢慢移向城郊,而工业区则位于河边。

奥古斯都别墅：酒店和花园在旧水塔里
Villa Augustus Dordrecht:
Hotel and garden in old watertower

多德雷赫特利尔公园：学校和公司创建工业园区
Leerpark Dordrecht:
Schools and companies create industrial campus

工业文化遗产再利用
E-using industrial heritage for cultural activities

Riedijks港口：现代公寓尽享河流的景色
Riedijkshaven Dordrecht: Contemporary
appartments with view on crossing of rivers

PlanTij工程：紧靠码头的水上住宅
Plan Tij : Intimate residential with Boating Docks

多德雷赫特 DORDRECHT　　WATER TOWNS

21世纪初，多德雷赫特的建筑类型可以分为三大类。第一类是大量毗邻城镇历史中心的邻水高品质公寓。沿河边的公寓或联排别墅越来越受欢迎，尤其是里戴克斯(Riedijks)港口附近新建的现代化公寓大楼。第二类是致力于提高景观品质的低层建筑。靠近城镇中心、位于废弃的水存储区域的Plan Tij工程就是一个很好的例子。这里的普通房屋都建在水面之上，而其中面积较大的房屋都带有船只停靠的码头。第三类是由工业遗产改造的建筑，例如由古老的发电厂翻新改造成的城市剧院和一家由旧水塔改造的精品酒店。所有这些例子都映衬了休闲经济的兴起。这是多德雷赫特转型背后的重要引擎。

最引人注目的还是多德雷赫特的"利尔公园"项目。项目的主体是一所技术专业学校校园。该校是一所为海事工业综合体输送高规格学生的专业学校。许多专业教育机构、当地公司和服务部门都集中落户在这里，以求增强公司和学生之间的直接联系，并促进知识在经济中的合理分配。这个项目采用了相对较新的发展模式，在多德雷赫特原本丰富的历史上添上了浓重的一笔。

编者感言

当一个城镇迈向辉煌，人们总是喜闻乐见的。但当要它脱去历史华丽的外衣，回归到平淡，是不是所有人都可以欣然接受呢？我们不喜欢用衰落这个词语，因为这往往是相对而言的。人们总是会用对比来评判事物的成功与失败。其实，人生最大的勇气并不是沉溺于成功，而是敢于直面惨淡的衰败。须知，这只是历史的潮起潮落，而在低谷兴许能收获受用一辈子的财富。

多德雷赫特承载了太多的历史，经历过绚烂繁华，但终究回归了平淡，并开拓了一条全新的发展之路。它所承载的历史足以在世界史上留名。经济的崛起，贸易当道，海上马车夫的故事在这里酝酿和爆发。人们在这里铸就梦想、寻找梦想，不同种族、来自各地的商人在这里汇聚。这完全是一个为了贸易而生的城镇。

可以看到，有那么多人在这里努力着。古今中外，穿越历史的400年，记忆仍然鲜明。人们在多德雷赫特满载而归，带着无尽的回味，乘着车船，驶向了下一个城镇。

阿尔梅勒地图
Map of Almere

阿尔梅勒 ALMERE　　　　WATER TOWNS

图片来源于 http://www.almerebeeld.nl/

阿尔梅勒城市天际线
Almere City skyline

广厦千万间，间间有个性——阿尔梅勒

与前面那些充满历史典故和古老建筑的小镇完全不同的是，阿尔梅勒作为一个新兴小镇的代表，缺的是历史文化积淀。它没有什么可以依托，但正是因为这种一穷二白的现实，反而让它无所畏惧。既然自新，那么就新颖到底。创新俨然成为了这个城镇的代名词。

阿尔梅勒这种敢于尝试和创新的魄力非常让人喜欢，尤其是居民直接参与房屋设计的这种互动方式形成的定制房屋，使得建筑形态美和功能特定性完美融合。现如今的许多住宅大都模式化，更多地被冠以开发商和设计师的想法。其实，这不是人们的理想家园。

实现美好生活，居者有其屋是一方面，其屋为其梦才是初心。试想，一个新的城镇让你实现了理想家园的梦想，而且还可以全程参与设计和施工，性价比高，谁能不心动！

阿尔梅勒是实现梦想的地方，不但是设计师的梦想，更是居民们的梦想。

正是，已得广厦千万间，小镇居民俱欢颜。原来，梦想就在眼前。

阿夫鲁戴克大堤
Deltaworks

阿夫鲁戴克大堤工程
Construction afsluitdijk

建造新的陆地
Makin of the new land

完工的阿夫鲁戴克大堤
Afsluitdijk after delivery

阿尔梅勒 ALMERE

WATER TOWNS

阿尔梅勒是为数不多的荷兰新兴小镇之一。它源于大海，平地起建，没有显著的文化历史，却居住着勇于挑战、善于创新的人们。长久以来，在须德海工程未建之前，阿尔梅勒一直处于海平面以下，从未有人在那里生活，没有文化积淀。如今，阿尔梅勒成为了一个拥有强烈的城镇意向和将近20万居民的活跃城镇，一个被建筑和新型文化围绕的中心地带。阿尔梅勒从零开始，一步步走来，直至发展成熟，已经铸就了自身的品牌。下一个飞跃性的机遇将在2022年兑现，荷兰世界园艺博览会的筹备建设正在有条不紊地进行中，这将为阿尔梅勒带来更多的国际关注。

阿尔梅勒的故事开始于一项巨大的荷兰国家工程，即历史上的须德海工程。在19世纪，为了使土地更安全，并创造出更好的农业条件，须德海工程应运而生。该项目由荷兰著名水利工程师莱利进行设计和总体规划。须德海工程包括由5个圩区和一座长约32公里的阿夫鲁戴克拦海大堤组成。大堤将须德海与外海隔开，将内湖变成了一个1100平方公里的淡水湖，称之为艾瑟尔湖。阿夫鲁戴克拦海大堤于20世纪30年代完成，圩区在60年代完成。在艾瑟尔湖内，最大的人造土地是莱沃兰圩区，在这块区域上规划了两个新兴城镇：莱利斯塔德和阿尔梅勒。两个新城镇都建在海平面以下约2米处。莱利斯塔德得名于须德海工程的总设计师莱利。阿尔梅勒则得名于古代德国对水的称呼，原意为"大湖"。

阿尔梅勒最初为城市溢流而形成的城郊镇，以此来应对战后人口的迅速增长，并同时解决阿姆斯特丹周围地区的住房短缺问题。阿尔梅勒的建设始于1976年。起初只是一些建造在沙地上的小型联排房屋，最早入驻这里的居民原本大多数也是出生于阿姆斯特丹。后来的几十年里，小镇迅速扩张。虽然莱利斯塔德比阿尔梅勒建设时间更早，并且规划建成的城镇规模比阿尔梅勒更大，然而由于阿尔梅勒在地理区位上更加靠近阿姆斯特丹和兰斯塔德，因而发展速度更快，反超了莱利斯塔德。

须德海工程
Zuiderzee Works

荷兰案例 Cases from the Netherlands

阿尔梅勒彩虹社区
Rainbow neighborhood

阿尔梅勒彩虹社区
Rainbow neighborhood

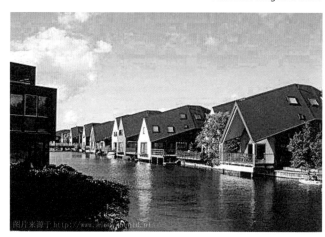

早期的阿尔梅勒住房设计案例
Early examples of almere housing by design

阿尔梅勒 ALMERE　　　　　　　　　　WATER TOWNS

阿尔梅勒最初被规划成一个多核城镇,一些居住区被分散设计在绿色景观之中。这种设计理念是为了让人与大自然充分接触,并提供悠闲惬意的户外生活。在规划第一阶段,将70%～80%的住宅规划为社会性住房,主要为梯田式住宅,多是拥有独立花园的小型公寓,可以欣赏到开阔的优美风景。规划者希望,阿尔梅勒可以成为60～70年代阿姆斯特丹地区城内房屋更新换代的选择之一。作为一个新城,新鲜是这里的特色。丰富的绿色空间和广阔的视野让人倍感惬意。然而,文化的缺乏和分散的城市配套服务,却成为小镇居民的遗憾。

随着阿尔梅勒的不断发展,政府和规划师逐渐意识到,必须扩大城镇的中心,为居民提供更紧凑的配套服务。多核发展的规划模式被重新审视,发展集中性的城镇和中心绿地被青睐。

为弥补历史文化缺失,使居民有存在感和归属感,规划师根据场地特征,为每个新的社区赋予了特殊的设计主题,从而提高了社区的识别性。通过建筑和城市设计形成阿尔梅勒特有的城市意象,使得阿尔梅勒俨然成为了一座时尚城市的代表,一所设计新风潮的建筑学院。比起任何其他荷兰小镇,阿尔梅勒更像是现代规划方法论的展示橱窗。这种主题设计方法被持续应用了近40年之久,赋予了一个原本没有历史渊源的小镇一种强烈的、令人印象深刻的特征。早期比较经典的案例应该是"彩虹社区"。"彩虹社区"的每组建筑颜色都以彩虹色为设计色调。另一个案例则是非常受欢迎的"电影社区"。"电影社区"的街道都以电影明星的名字命名,并且每5～10栋建筑为一组的建筑群都由不同的建筑设计师主持设计。阿尔梅勒的扩展和发展已成为荷兰乃至世界规划和建筑的试验展览场所。在这里,居民对于城镇规划和建筑设计拥有发言权。如今,阿尔梅勒已是DIY城市发展和住宅建设的市场领导者,为居民设计和建造自己家园提供了有趣的机会和尝试。

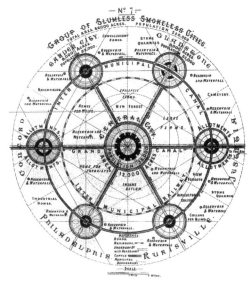

埃比尼泽·霍华德:多核城市规划蓝图
Ebenezer Howard : Blueprint of polynucleair towpn planning

阿尔梅勒新市中心
Almere new Towncenter

雷姆·库哈斯(OMA)的市中心总体规划
Masterplan Towncenter by Rem Koolhaas OMA

克里斯蒂安·德·波特赞姆巴克设计街区
Kristian de Porter Zamba Barker Design blocks

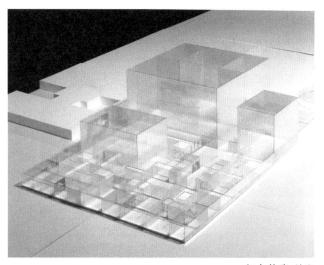

阿尔梅勒剧院
Almere Theatre

阿尔梅勒　ALMERE　　　　　WATER TOWNS

阿尔梅勒在发展初期被定义为阿姆斯特丹的配套城镇。居民希望生活在拥有良好的自然环境和与主城有着便捷交通连接的地区。但随着居民人数的迅速增长和地方企业开放性的不断增强，阿尔梅勒从一个依附于主城的通勤城镇逐步发展成为一个综合城镇。这就需要在安置居住和提供工作之间有一个良性的平衡。当地政府意识到，让阿尔梅勒发展成为一座配套成熟的城镇，除了营造优美的居住环境外，还要达到产业发展，配套齐全，交通便捷，社区功能完善。

政府还意识到，要想将阿尔梅勒打造成新型市镇典范，形成独特的品牌，首先需要一个良好的规划。经过了激烈的规划设计竞赛，著名建筑师雷姆•库哈斯(Rem Koolhaas)的规划方案获得了认可。他的设计奇妙有趣，在市中心引入了人工山，这座人工山小有起伏，公共交通线路穿山而过，巧妙地将城市停车场以及公共交通线等公共设施覆盖起来。起伏地形打破了原有平坦的小镇地貌，让人耳目一新。为了使设计变得更加准确和受欢迎，库哈斯也参考比对了荷兰老城中心的各种案例，设计将所有的建筑分割为较小的区块，呈颗粒状分布。设计手法不是用古典建筑风格，而是采用当代设计风格，使得新城镇中心，在视觉上显得更加活泼、不同寻常。

除了库哈斯的规划方案，阿尔梅勒还有许多其他设计亮点。小镇中心的一处街区就是由著名的法国建筑规划先锋设计师克里斯蒂安•德•波特赞姆巴克(Christian de Portzamparc)设计的。这里居住、商业以及配套都十分完善。另一处设计亮点则来自著名的日本SANAA事务所，沿中心河设计的小镇歌剧院以及社区中心项目也十分抢眼。这些由国际著名设计师领衔的设计作品给阿尔梅勒创造了一种独具特色的气质特征。虽然评论家认为阿尔梅勒居住的中产阶级缺乏文化底蕴，但人们依然可以看到这些特殊建筑给城镇带来的品牌效应。

受这些独具创意的公共项目的影响，阿尔梅勒的住宅开发项目也充满了革命性的气息。在Homerus社区新建的3000套房屋的项目中，业主与开发商共同参与了房屋的设计、融资与建设。在此之前，住房一向是由专业开发商策划，居民的参与活动只能在他们签署合同并获得钥匙后开始。这种革命性的房屋定制化模式创造了一种城市发展的新方式。

Homerus 社区
Homerus neighborhood

人们建造自己的房子
People built their own houses

2022年阿尔梅勒荷兰国际园艺博览会由威尼·马斯(MVRDV)设计
Floriade 2022 designed by MVRDV/Winy Maas

阿尔梅勒 ALMERE　　WATER TOWNS

基于丰富的土地和生态资源，阿尔梅勒的居民人数预计会增长至35万左右，届时将会成为荷兰主要的大城镇之一。然而实现这一目标还需要十几年的时间。目前正在建设的最新社区是阿尔梅勒森林(Almere Hout)，所有传统的开发和建设方式都被赋予了创新思维。土地将作为一个开放的资源，政府在确定总体规划的前提下，允许业主和开发商适度自由发挥。这种总体控制、单元自由的规划方式在荷兰乃至世界的城镇建设中都是标新立异的，为新城镇的规划发展闯出了一种新模式。

阿尔梅勒成功的申办了2022年荷兰世界园艺博览会。这一盛会将会极大的促进荷兰园艺产业的发展，也使阿尔梅勒从中获益。阿尔梅勒政府已经邀请著名荷兰设计公司MVRDV的威尼·马斯(Winy Maas)设计师为这次活动提供创新、可持续性的总体规划。此项总体规划设计了一个覆盖公路系统的强大的网格，以此来加强城市现有的基础设施。会后，展览场地将成为阿尔梅勒城镇一个具有混合功能的区域，与其他城镇空间联系融合，开辟阿尔梅勒城镇未来发展的新纪元。

编者感言

阿尔梅勒是颇具灵魂的城镇。虽然很新，也被一些人说没有文化和内含，但阿尔梅勒的这种无畏，也许正是因为没有历史的包袱，所以可以大胆创新，并成为一群创新者和理想者的梦想城镇。反观中国，许多城镇都有悠久的历史，但这些历史和文化有的时候也成为了包袱，让城镇观望创新，不能与时俱进，不能很好地让古今融合。经常看到这些城镇在矛盾中不停摇摆。一方面，说着自己是有历史有文化的；另一方面，却也不知所以地羡慕许多现代化城镇。在到底是做文化城镇还是追求现代化的选择中摇摆不定，以至于到最后丢掉了自己真正的文化和特色，去追求所谓的繁华和都市氛围。

城市的灵魂和核心提炼是很重要的。什么都想要，往往结果是什么都不像了。艺术城镇就是艺术围城，文化城镇就主打文化。为什么都要通通都要成为摩登城镇呢？这样的感觉就像是硬生生要在牛津大学城建一个迪拜最高楼一样，而代价是拆毁牛津的百年老楼，实在让人为之扼腕。别说只有几十年历史的城镇没有灵魂，阿尔梅勒的灵魂就提炼得很好。

创新，一往无前，让设计师尽情发挥，也让平凡的人得以圆梦。只要这样，它才完整，才吸引人。

2022年阿尔梅勒荷兰国际园艺博览会
Floriade 2022 planning

艺术小镇
ART TOWNS

来自荷兰的案例
Cases from the Netherlands

76/贝亨　　BERGEN

84/拉伦　　LAREN

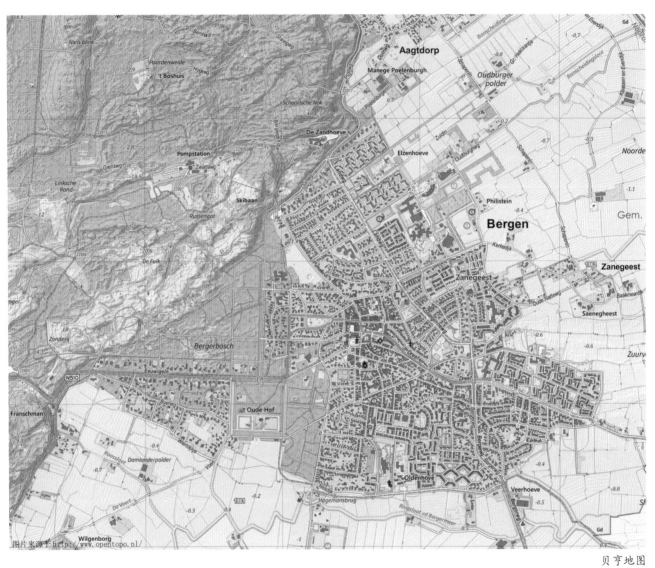

贝亨地图
Map of Bergen

贝亨 BERGEN

ART TOWNS

贝亨"废墟教堂"
"Ruinekerk Church" Bergen

上帝、历史和艺术家共用的画布——贝亨

会让那么多艺术家趋之若鹜，并在其定居的小镇，一定有它的迷人之处。这小镇给人印象最深刻的就是它漫长而一望无际的海岸线。这不是轻易可以拥有的景致。幸运地躲过大洪水的贝亨的确有着传奇色彩。说它被上帝眷顾，的确不假。吊灯的血色撑起了神迹，唤来了越来越多的朝圣者。但是，如若上帝眷顾，为什么它又那么多次被卷入战火。圣洁和神圣的教堂也在战火中被焚毁，如今空有骨架，和澳门的大三巴教堂遗址有一些类似。

好在贝亨的奇迹没有终结。火车为贝亨带来了一大批内心无处安放的艺术家。当这群艺术家遇到贝亨，那教堂染血的吊灯似是为他们指明了方向。这样一个浪漫富有传奇色彩的小镇，比阿姆斯特丹更符合他们心中的理想国。艺术自此成为了贝亨的灵魂。艺术家群体和沙龙迅速地形成，并创立了学校。

所以，有的时候经历磨难，也能让小镇披上一丝独有的气质。也许吸引这些艺术家的正是这传奇般的故事和沧桑吧。

中央道路的历史照片
Historic image of central road

废墟教堂
The Ruinekerk church

教堂边的房屋
The house next to the church

教堂边的房屋
The house next to the church

贝亨 BERGEN

ART TOWNS

贝亨是20世纪荷兰最著名的艺术小镇之一，这多少是为有很多著名艺术家都曾就读于贝亨的Bergense学校。尽管这里拥有悠久、感人的历史，却也是三场战役的战场。贝亨的最特别之处是其位于沙丘到圩田的过渡地区，有着逐渐变化的景观和植被。这为贝亨后来的发展和定位创造了得天独厚的优势。就如我们今天所看到的，贝亨已经成为著名的艺术小镇和海滩度假村。许多表现主义建筑的别墅都成为了贝亨小镇的艺术遗产。

贝亨坐落于北荷兰省(Noord-Holland)，其历史可以追溯到800年前。由于其位于北海沿岸沙丘和北荷兰绿色圩田的交叉口，此地不同类型的景观造就了贝亨的独特之处，泥炭土变成砂地，慢慢形成了丰富多样的地表植被、树木和地形。最古老、最富有的部分位于沙丘附近，因为土地高，所以可以免受洪水和风暴的侵害，其植被为当地居民提供了舒适的生活环境。在早期，北海和须德海之间总是有洪水隐患。1421年发生了"圣马丁伊丽莎白大洪水"(Sint Elisabeths Flood)。荷兰的版图被这场突如其来的灾难打乱，贝亨邻近的佩滕小镇完全被洪水冲毁了。

荷兰在17世纪"黄金世纪"时是世界上最成功的国家之一。贝亨得益于Bergermeer圩田，保证了其水利安全。城镇收入则主要来自于海外贸易。贝亨因为距离16～17世纪荷兰共和国的两个经典城镇——阿姆斯特丹与阿尔克马尔较近，所以或多或少卷入了与之相关的历史事件。

1574年，贝亨遭受了阿尔克尔玛周边地区与西班牙之长达80年的战争的波及。在1799年贝亨也成为了战场，当时英国和俄国军队在巴达维亚(Bataafse)共和国期间，试图从荷兰放逐法国统治者，恢复奥兰治王朝。因此，在贝亨有一座纪念碑，为了纪念远离祖国、死于战争的1700多名俄罗斯士兵。但是，贝亨作为战场的角色并没有就此结束。荷兰北部的海岸线在第二次世界大战中具有战略性意义。那时德国在贝亨附近的海岸线增强军事设施，成为德意志帝国的东部防线"大西洋墙"的一部分，周边地区则作为德军的一个军事机场。

此外，围绕贝亨的一个重要故事是关于"贝亨的奇迹"。1421年大洪水灾害过后，贝亨人民发现邻近的佩滕小镇的所有教堂都被海水吞没了。出于尊重和信仰，教堂文物被拾起和恢复，吊灯里的水也被保留了下来。6个月后当牧师检查吊灯，原来的海水竟变成了血水。在那个时代，这被视为"圣灵的介入"。贝亨因此成了被人们敬拜的场所。来自荷兰和海外的许多重要人物都会去贝亨朝拜。为了容纳这一大批的崇拜者，突出贝亨新的宗教地位，政府在已有的教堂旁建了一座更大的教堂，这两座教堂一起被称为"朝圣教堂"，以此纪念所有来到贝亨的朝圣者。然而，由于突袭阿尔克尔玛战争的负面影响，教堂在16个世纪成为西班牙侵略者的目标，部分被毁坏，只剩下骨架。如今，这个骨架仍然矗立在小镇的中心，并有个不祥的名字叫"废墟教堂"。

荷兰艺术家查莉·托洛普
Dutch painter Charley Toorop

贝亨艺术中心
Art center Bergen

贝亨艺术中心
Art center Bergen

贝亨街道
Bergen street

贝亨 BERGEN

ART TOWNS

直到19世纪末,贝亨才达到巅峰。许多重要的事情几乎同时发生在这个小镇中。作为19世纪后期浪漫主义运动的产物,一些小规模社区在浪漫主义文学中被理想化。当时,这样的现象发生在很多大城市中。同时,蒸汽火车和蒸汽电车等新技术的到来,实现了人们远行的愿望。1901年,贝亨有了直达阿姆斯特丹的火车。至此,那些充满好奇而又有文化意识的阿姆斯特丹人逐渐到访贝亨,特别是很多艺术家选择在贝亨定居,慢慢的一个名为Bergen Binnen的小区诞生了。这是贝亨作为艺术小镇的起源。贝亨的主要特色之一是矗立在沙地上的茂密树木和独立的海边村庄。海滩旅游始于19世纪末,有早期的旅游设施,如小旅馆。

在二十世纪早期,艺术家经常围绕艺术理念和风格聚集在一起,那些理念、风格让他们在国内外都享有盛誉。Bergense学校关注表现主义和形象艺术,有成名的中心人物,也有其他许多贡献者。学校的创始人是法国艺术家亨利•拉•福科尼耶(Henri le Fauconnier),他从法国带来了他的艺术理论。但最著名的艺术家应该是画家查莉•托洛普(Charlie Toorop)和雕塑家约翰•雷德克(John Readecker)。雷德克是荷兰最著名的雕塑家,纪念二战死难者的阿姆斯特丹大坝纪念碑就是他的作品。托洛普和雷德克的作品具有极高的艺术价值,被收藏在许多荷兰顶级博物馆中。阿尔克马尔现代艺术博物馆收藏了大量来自Bergense学校的作品。

来贝亨居住的艺术家中也包括从阿姆斯特丹过来的建筑师。这一时期,在法国建筑师勒•柯布西耶(Le Corbusier)提出现代建筑之前,荷兰已经形成了一种表现主义艺术形式,被称之为"阿姆斯特丹学派"。随后,建筑师和追随者们开始在贝亨的废墟教堂附近和海滩沿路建造房屋和别墅。多年来,这些别墅区不断扩大,现在已经成为荷兰最美丽、最受尊敬的别墅区。

查莉•托洛普自画像
Charley Toorop Self-portrait

建筑师斯塔尔设计的"树皮别墅"
Villa de Bark Architect Staal

建筑师 Blaauw 设计的"Meerhoek别墅"
Villa Meerhoek Architect Blaauw

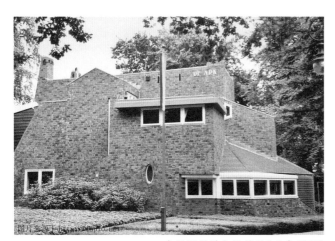

建筑师斯塔尔设计的"方舟别墅"
Villa de Ark Architect Staal

建筑师 Kropholler 设计的"Mezennest别墅"
Villa Mezennest Architect Kropholler

贝亨 BERGEN

ART TOWNS

最著名的别墅区湖区(Meerwijk)，集中有16栋别墅，是1916年阿姆斯特丹瓷砖商人海斯蒂(Heystee)委托设计和建造的。他让著名的阿姆斯特丹建筑师斯塔尔(J.F.Staal)提出总体规划方案，设计别墅建筑。唯一的附加条件是别墅需要应用海斯蒂自己公司的瓷砖。斯塔尔一共设计了5套别墅，并带领了一群阿姆斯特丹的建筑师设计了其余的11套别墅。在这些建筑师中，有克罗弗勒(A.J.Kropholler)和皮特•克拉默(Piet Kramer)等著名人物。克拉默是阿姆斯特丹学派的三大建筑师之一。克拉默、德克勒克(De Klerk)和范德梅基(Van der Meij)一起设计了阿姆斯特丹最著名的建筑之一——航运大厦(Scheepvaarthuis)。后来，克拉默成了阿姆斯特丹建设部的首席设计师，设计了阿姆斯特丹二十世纪诸多著名的桥梁。不过，湖区最重要、最著名的别墅是由建筑师斯塔尔自己设计的，比如"方舟别墅"和"树皮别墅"等。在后来的几年里，克拉默在贝亨的作品不幸丢失了。别墅区后来也成了建筑展示的场所。甘蔗屋顶、弯曲的砖和木制基底从此大受欢迎，荷兰其他地区也纷纷效仿。

如今，贝亨是一个典型的荷兰度假海滩休闲和文化旅游目的地。它的建筑和景观成为了国家古迹。到贝亨旅游的人一定会被当地的氛围所感染。沙丘、森林、圩田和美妙的建筑遗产都证明了贝亨作为艺术之城的特殊地位。

编者感言

历史千百年，辉煌总是难以永恒。但这些历史和故事都是时间遗留给小镇最好的礼物。上帝关上一扇门的时候，也会为你打开一扇窗。这句话不假，并且在荷兰一次次地被证实。历史是华丽的外衣，也有丰富的遗产。但时间是向前的，我们必须根据现实制定新的计划，找到新的目标。贝亨没有重建"废墟教堂"，有着一定的历史意义。苦难也许让贝亨更加神秘和风情万种。艺术家往往是通灵的，他们对自然和周遭的事物尤其敏感。也许贝亨真的有着他们要找的神迹和圣洁，他们在到达的那一刻便明白了。一为单，二为伴，三五七成师，四六八凑对，过十成群。当一群富有灵气的人聚集起来，火花和艺术便在此萌芽和爆发。

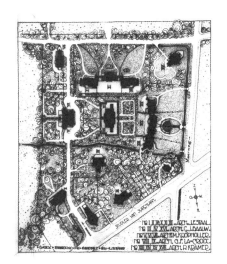

Meerwijk别墅区总体规划
Masterplan Villapark Meerwijk

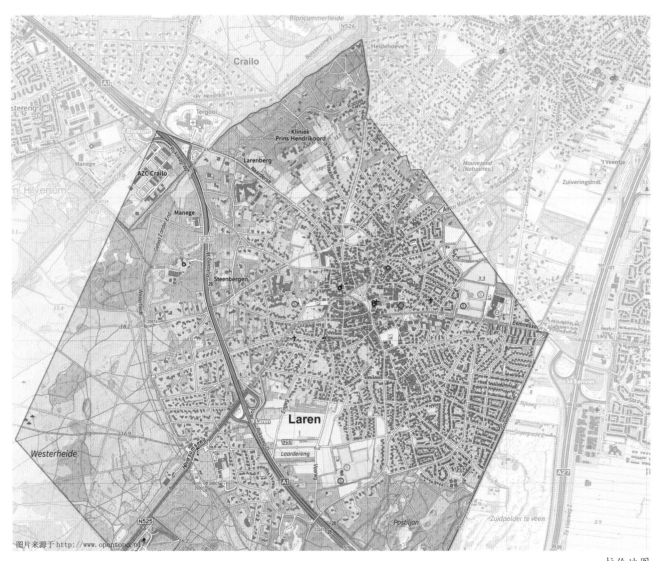

拉伦地图
Map of Laren

拉伦 LAREN ART TOWN

ART TOWNS

拉伦镇中心
Laren town centre

千古田园，一曲牧歌——拉伦

作为荷兰为数不多的丘陵小镇，拉伦有着得天独厚的景观资源。起初，作为开放的村庄，人们共用着农场，过着悠闲的田园般的生活。这样的地方的确很适合艺术创作，让人摒弃了许多世俗的杂念。牧场、丘陵、童话般的农舍，一切都符合艺术家们接近自然的那种诉求。和贝亨类似的是，拉伦也因为现代交通工具的建成而成为艺术家流连的处所。

坐上电车，远离喧嚣，40分钟的不远处就是心灵的故乡。那么，住过去？何乐而不为呢？更何况那时候没有什么高房价。可以说，艺术家成就了拉伦，但也可以说，拉伦成就了艺术家。

有的时候，心跳的地方才是故乡。而有的时候，让我们心灵归复平静的地方，才是故乡。

历史景观分析
Historic landscape analysis

拉伦大教堂：圣约翰大教堂大教堂
Laren, basilica: De Sint Jansbasiliek

拉伦镇中心
Laren town centre

拉伦 LAREN ART TOWN

ART TOWNS

　　拉伦是建立在丘陵上的一个小镇，这在荷兰是比较少见的。小镇最初是一个孤立的农业社区，慢慢蜕变成19世纪后期的艺术家社区。优美的自然环境和典型的农场建筑被证明是非常有吸引力的。著名的画家梵高和蒙德里安也在拉伦停留了不少时间。1883年蒸汽电车的到来给艺术家的繁荣提供了更多的便利，也使拉伦成为一个早期的通勤小镇。而如今，丰富的文化遗产和迷人的小镇中心以及高级的森林别墅，使拉伦成为了许多名人和新贵的向往之地。

　　在荷兰中部，拥有一处史前时代的天然景观——乌特勒支丘陵。它是最后一个冰川时期遗留下来的，来自北极地区的大量冰川重塑了荷兰的平原地形。这条位于艾瑟尔湖（Ijsselmeer）和莱茵河之间的含沙量不高的山丘群一直绵延50多公里。乌特勒支丘陵，实质上是冰川堆石，总共包含约30个山丘，山丘高度在冰川时期的海平面以上20～50米之间。乌特勒支丘陵很容易通过森林和荒野的植被以及地形来识别。这与地形平坦多是泥炭地质的荷兰水国景观大相径庭。历史上，该地区还形成了一种与荷兰其他城镇如莱顿和代尔夫特不同的社区模式。这里的村庄大多是以农业生产和养牛为经济核心的森林社区。

　　拉伦就是这些社区的其中之一，位于乌特勒支丘陵的北部边缘，靠近艾瑟尔湖。非常有趣的一点是如今拉伦极具吸引力的丘陵地形和迷人的森林景观可以追溯到很久以前的地质过程。也因此很长一段时间内拉伦只是处于荒原和森林之间的一个边缘村庄，有着独特的生存方式和地方文化。起初拉伦只有不超过50个农户，这种情况持续了很久。该地区的一个特别之处是所有农户共用土地，这种没有私人农田的状况意味着整个社区是开放的，不设门和栅栏。这种共用土地的农业方式决定了村庄的布局。人们会在所有农场之间设置一个开放空间。小镇的名称拉伦来自荷兰语词汇Laar，意思是森林里的开放空间。每天早晨和晚上，牲畜将被牵引到村庄的一个大的开放空间。随着时间的推移和村庄的不断扩大，中间的开放空间被越来越多的农场包围，最终成为村庄的中心地。为了形成护荫区域并同时生产一些木材，人们在被称为"es"的这个开放空间种植了一些树木。后期这个开放空间又被用作每周定期开放的市场来销售当地生产的商品。这种具有大开放空间且中间有大树的村庄类型在荷兰被称为"开放的村庄"。

1885年拉伦历史照片
HIstoric photo 1885

安东·莫夫画作中的拉伦景观
Landscape Laren painting by Mauve

拉伦街道景色
Laren street scenery

拉伦建筑
Laren architecture

拉伦 LAREN ART TOWN

ART TOWNS

很多世纪以来，拉伦一直保留着乡村社区的特色，未受外界影响，形成了本地独特的语言，独特的着装和农场建筑风格。从当时的照片可以看出这是一个美丽和谐、乌托邦一般的村庄。这里的男女着装独具特色，农场建筑风格也各式各样。农场建筑的长度通常超过25米，前面部分是带宽敞客厅的起居区，面积较大的后半部分则供牲畜使用，两个区域中间被设有烟囱的厚砖墙分开，烟囱和暖气及厨房区域连通。农场屋顶上铺的多是甘蔗，隔热效果非常好，使屋内冬暖夏凉。对农场来说，最重要的是能够减少秋季强暴风带来的影响。风暴期间牲畜会从农场南边的门被引导进入牲厩。随着时间的推移，木制结构的农场长屋顶因为重力稍稍下沉，屋顶的形状开始弯曲，中间成为最低点。这样的变化过程反而带给人们的一种新鲜感，甚至觉得很是优美。很快美丽如画的建筑受到了来自许多村外游客的青睐。

19世纪，著名的荷兰现实主义画家约瑟夫•伊斯拉埃尔斯(Josef Israels)发现了拉伦，被它的美丽所感动。后来许多艺术家朋友追随他的脚步来拉伦居住。他们买下老农场，然后改建成新别墅。一个艺术家社区也因此逐渐形成。那个时代最著名的艺术家之一安东•莫夫(Anton Mauve)也搬到了拉伦居住。他的到来又吸引了新的追随者。莫夫也和梵高相识，在梵高年轻时的艺术形成期，他曾帮助过梵高，传授了新的技能，给这位有天赋的年轻人指引了新的艺术方向。著名的美国艺术收藏家辛格(Singer)也爱上了这并搬来居住。如今非常出名的辛格现代艺术博物馆就是这位美国艺术家给拉伦留下的历史文化遗产。另外，20世纪最著名的画家之一蒙德里安(Mondrian)年轻时也曾在拉伦居住。

拉伦成为19世纪后期艺术家聚集地的一个重要原因是现代化基础设施的建成。1883年连接阿姆斯特丹和希尔弗瑟姆之间的蒸汽电车在拉伦设了一个停靠站，从此阿姆斯特丹到拉伦的时间都不用40分钟。这意味着拉伦交通领域的重大变革，在此之前两地的交通基本靠马车和船只，这两种交通工具自然没有电车快。至于公路和汽车，那是60年以后的事了。

拉伦不仅仅因其迷人的风景成为一个艺术城镇，在更广泛的意义上也作为一个通勤城镇和休闲区更好地服务于艺术家和居民。拉伦居民对艺术的极大兴趣以及相对富裕的经济状况都能从许多建筑和遗迹上清楚地捕捉到。拉伦有数以百计的国家和地方建筑文物。除了农场以外，这些文物也包括二十世纪上半期的当代设计风格的建筑。拉伦本地的著名建筑设计师Wouter Hammdorff不仅设计了许多工作室用途的别墅，还设计了辛格现代艺术博物馆，这是一个极其容易识别的具有浪漫主义风格的建筑，将传统的造型和材料用现代的方法完美地诠释，是传统和现代的完美结合。他设计的别墅从20年代开始就是许多电视剧的取景地。其作品中另外成为景点的是拉伦水塔，它结合当地森林和地形的环境，将优雅外形和水塔功能的完美融合在一起。

拉伦辛格现代艺术博物馆
Singer museum Laren

乌特勒霍伊风景
Het Gooi scenery

建筑师Hamdorff设计的拉伦水塔
Water tower By architect Hamdorff

建筑师Hamdorff设计的别墅
Villa Architecture by Architect Hamdorff

拉伦 LAREN ART TOWN　　　　ART TOWNS

拉伦是乌特勒霍伊(Het Gooi)地区的中心。这个地区的其他城镇，如布拉克里姆(Blaricum)和希尔弗瑟姆(Hilversum)，都是荷兰著名的传媒产业城镇。许多著名的电视人以及来自阿姆斯特丹地区的有影响力的名人也都喜欢选择居住在拉伦及其周边。这些富人选择在乌特勒霍伊地区生活，就是因为这里的别墅具有独特的高树及大草坪景观。此区域也有许多餐馆，画廊，精品店和其他服务设施。由于现今居住的这些特定人群，拉伦成为了荷兰著名的富裕城镇。在小镇中，另一个特别设施是"罗莎•斯佩尔公寓"，这是一个艺术家和科学家的老年之家，可以在一个艺术和生产服务的环境中老有所养。这是艺术小镇拉伦实行的一系列独特的配套举措之一，深受专业人士的欢迎。

在荷兰，许多具有悠久历史的城镇在功能运用上已经成功转型，同时也保留了其美丽的自然和文化遗产。拉伦就是这样成功的例子：从农业到艺术，从森林农庄到艺术新贵的追捧热点。所有这一切都归功于经久不衰的景观和对文化遗产的热忱保护。

编者感言

19世纪这些艺术家的追求和开拓真的很让人欣赏。喜欢一个地方，就会去追逐；也许找到这个地方需要旅行许久，但一旦选择了就会义无反顾。也许艺术就是如此，需要和志趣相投的人在一起，需要有师长，有同伴，有指引，有陪伴，有鼓励，有交谈。艺术家大多是孤独的，那么，让他们聚集起来的地方的确是该如伊甸园一般美好。现在的年轻人总很渴望的一样东西，就是一个舞台。这个舞台其实涵盖很多，是平台，是认可自己的同伴，更是能擦出火花的交际圈。

所以，作为一个城镇，怎样吸引这些艺术家、创作者，是十分重要的。往往需要给他们创造的不仅仅是口号和所谓的政策，更要凸显整个城镇的吸引力，将自然和人文景观优美之处开放给他们，让他们驻足不舍离去。只有留恋才能真正地让他们可以安心地在这里生活。所谓的安居乐业，激发灵感，才是艺术家最纯朴的追求吧。

产业小镇
PRODUCT TOWNS

来自荷兰的案例
Cases from the Netherlands

94/	利瑟	LISSE
106/	阿尔克马尔	ALKMAAR
114/	代尔夫特	DELFT
124/	恩斯赫德	ENSCHEDE

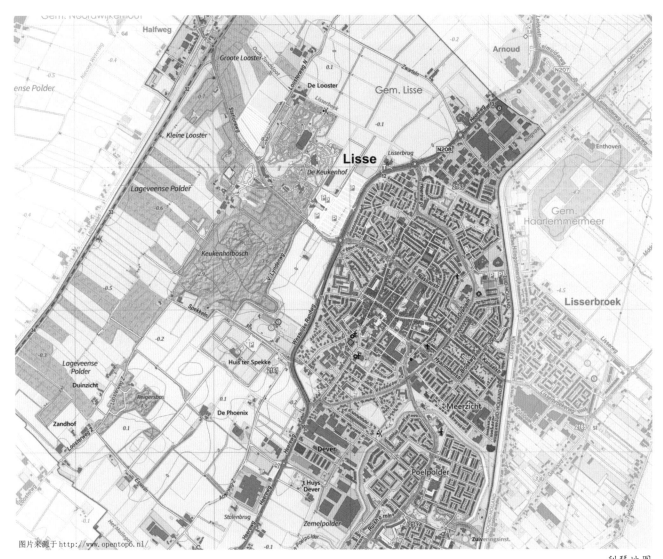

利瑟地图
Map of Lisse

利瑟 LISSE　　　PRODUCT TOWNS

利瑟街景
Lisse street view

一座守护着国花的小镇——利瑟

纵览那么多小镇，利瑟的故事就似那郁金香花海一般，让笔者觉得最明艳动人。相信，没有人是可以拒绝鲜花的，哪怕那些对花粉过敏的人，他们依然可以看着鲜花的照片，说上一句，好美！利瑟让我见识到了商业和资本的迷人之处，却也是因为人们对资本和利益的贪婪和追逐，让多少人一夜破产，让荷兰的经济一蹶不振。说实话，经济学的许多里程碑式进展都是源于荷兰。郁金香的奇迹和危机，让我们很好的看清了资本这把双刃剑。物极必反，丧失理性，必将付出相应的代价。但利瑟花匠的坚守却让我动容。他们把错误都归咎给了自己和人类，仍然深深地爱着郁金香，并为人们犯下的错误潜心忏悔，也不停钻研和优化球根花卉的种植。

这种面对逆境的不忘初心，让我看到了人类最原始的那份坚韧和执着。美丽的事物终究是美丽的，也许改变的只是人心。坚持善良的人心，坚守自己那份最初的追求，只要是纯洁的、虔诚的，就一路走下去，就算路上突然山洪淹路，就算月光不再明亮，前途一片昏暗，也要坚定地走下去。因为他们的手上有着一盆盆花卉，对着花卉微笑，那娇艳的植株迸发的花火终会照亮前途。

库肯霍夫公园
Keukenhof park

利瑟风车
Windmill in Lisse

利瑟花田
Lisse flower field

利瑟 LISSE

PRODUCT TOWNS

春天到访过荷兰的人，多半会去利瑟，甚至可以说，一定要去利瑟小镇。因为，那是世界上最美的花卉小镇，那里守护着荷兰国花，是传承和发扬着荷兰郁金香历史和荣耀的小镇，那里还用郁金香创建了一个举世闻名、让人类享有天堂般视觉体验的公园——库肯霍夫郁金香公园。

利瑟位于荷兰西部，隶属南荷兰省，历史悠久，记载在册的历史已超过了800年，当然也有可能更久。根据地名专家猜测，利瑟很可能意为栅栏或栏杆，小镇中心广场取名为"四方形"很可能也是源于此。这跟欧洲凯尔特时期的"四方形堡垒"的城镇特点有关，凯尔特人会在地势较高的地方或沿着水路搭建四方形小镇作为商品储存交易的场所和牲畜保护的场所，现在一些法国和英格兰的地名里仍含有利瑟就是山丘堡垒的意思；另一种说法是利瑟的意思是未固定的松土。

中世纪时期利瑟还非常小，1500年前后才有50户居民，连续不断的战争导致了此地的贫穷，当时人们仅靠捎泥炭为生。如今利瑟人口约2.2万，占地面积16平方公里。以郁金香为主的种植生产、花木贸易为主要产业，并依托于举世闻名的郁金香公园促进了旅游、服务业的发展与繁荣，成为荷兰美丽而充满活力的小镇，向世界持续传达荷兰国花的精神境界。

郁金香——上帝的礼物

1593年，第一朵郁金香由土耳其传入这个国家，被荷兰园艺学家栽培成功，这个美妙的杯状花朵"艳丽的让人睁不开眼睛，完美的让人透不过气来"，这个"上帝的礼物"让整个荷兰为之狂欢，如痴如醉。

自此开始，少女们对皇冠、宝剑和金子等作为信物都不再钟情，却要向花神祷告，把皇冠变鲜花，宝剑变绿叶，金子变球根，合起来便成了一朵郁金香。郁金香在当时不仅成了美丽、浪漫爱情的象征，还是情侣们钟情的恋爱信物。诗人和画家对这种来自异域的奇花也进行了无尽地赞美和书画。名流显贵们争相追抢，郁金香也成为了富有与品味的象征。

与浪漫的法国人不同的是，荷兰人还以冒险精神和商业头脑著称。在荷兰人中也不乏一些商业的投机者，他们在欣赏郁金香浪漫和美丽的同时，很快发现了其巨大的商业潜值。投机者不惜为郁金香编写浪漫的传说故事，还编织了嘲笑英国人不懂郁金香的土里土气的故事，刺激人们的欲望；利用荷兰发达的金融交易市场和商人们敢于冒险的性格，将郁金香的价格在市场中多轮哄抬，最夸张的价格甚至达到了每粒种子10000荷兰盾。荷兰一个叫霍伦的小镇中心至今有还保留着一栋大房子，门前竖立着三座郁金香的石头雕塑，就是为纪念这座房子是1633年夏天用三株稀有的郁金香换来的。除了后来被改建为天主教堂之外，这栋建筑本身并没什么特别之处，但是这里却是郁金香狂热的见证。

郁金香这个"上帝的礼物",在荷兰不仅绽放出了她的美丽、优雅与品味,让很多如利瑟小镇的人们对之尊敬与神往;也为当代经济产品的金融化发展奠定了基础;同时还让荷兰的商业投机者们谱写了一段"狂人舞曲",上演了一出让世界震惊的经济悲喜剧。

为失信而忏悔

在1636年至1637间的郁金香球茎贸易交割过程中,荷兰人创造了"期货选择权",且推出交易杠杆,允许买空。而今我们耳熟能详的"看多""看空""期权""期货"等证券市场名词基本都是那个时候由荷兰人创造出来的。1637年2月4日这一天,买方出现了大量的抛售,市场瞬间陷入恐慌状态,郁金香球根的价格暴跌。随后,签署的郁金香合同变得一文不值,持有郁金香合同的人开始逐个向前一个卖主追讨货款。昔日的景气场面顿时变成了凄风苦雨和逼债逃债的地狱。荷兰政府除了决定禁止郁金香合同签订,禁止投机式的郁金香交易外,也无能为力。一年之后,政府再次规定,允许郁金香的最终买主在支付合同价格的3.5%后终止合同,在这个打击之下,荷兰的郁金香投机市场一蹶不振。历史学家安娜·戈登加(Anne Goldgar)在《郁金香狂热》中写到:郁金香投机最大的破坏是针对荷兰人的价值观,有什么比欺骗和伪造更糟糕?让我们把自己全都当成骗子吧!郁金香贸易不是错在金钱,也不是错在商业,而是错在将建立在信任之上的有序社会放置一旁。它不仅让荷兰商人阶层在经济上被击垮,商业的失信也给社会带来了更加致命的影响,摧毁了整个荷兰经济,也让荷兰随后失去了经济霸主地位。

利瑟小镇的人们虽然对郁金香的遭遇感到愕然与惋惜,但他们并未因此改变对郁金香的欣赏、尊敬与神往,也未因郁金香市场的萧条而放弃与逃离。他们的行动不是停留在留恋与守望之中,而是选择种植其它如风信子等球根花卉,还不远千里从墨西哥引进孤挺花(星花属)和大丽花,从南非引进小苍兰和百合进行种植生产,以此来保持对郁金香等球根花卉生产的延续与发展。

利瑟的人们义无反顾的坚守着,不仅是因为小镇的人们对郁金香的喜好,更像是小镇人对上帝忠贞的一种信念。他们相信最美丽鲜艳的郁金香,无疑属于最深处的极乐花园的神物,是上帝恩赐给人间的礼物。他们不仅需要守护好这份天堂之物,还需要为市场失信导致对郁金香神物的亵渎而忏悔。

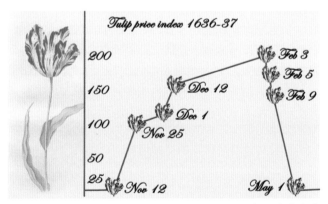

1636～1637年郁金香泡沫
Tulip Bubble 1636~1637

利瑟 LISSE PRODUCT TOWNS

当地适宜的土壤气候条件，还有那份着坚持与执着，磨练了小镇人们种植郁金香等球根花卉的高超技艺。人们用钻研与创新，开发出越来越多的具有不同形态和颜色的新品种。利瑟小镇的花匠用忠实与行动，守护着郁金香为代表的球根花卉，期待着郁金香重新焕发生机。

用郁金香为荷兰（上帝）守信

19世纪后，郁金香的生产贸易恢复了正常。利瑟小镇人们不忘前车之鉴，他们把守信放在了郁金香贸易的商业和利益价值之上。

利瑟花匠们重视生产与栽培技术的研发，让郁金香贸易在数量和质量上践行守信。利瑟小镇的人们与荷兰的球根种植的花匠一起开展了郁金香等球根花卉的生产技术研究。为了保障种球栽培质量的稳定，利瑟的人们利用蒸汽机技术，率先在郁金香种植区建立了水泵系统，以便控制地下水位的一致，保持球根大田生产质量的稳定；也根据郁金香各品种的品性特点，制定开花期的择心掐头技术，刺激种球生长、积累养分和降低病虫害；还建立高温和低温库，研发种球存储技术，满足郁金香和其他球根在一年中的存放质量和四季发运能力，确保郁金香的质量和供货时间上的诚信。

利瑟用完善的机制保障贸易合同的签订和守约。为避免郁金香狂热期间各种非专业的投机者介入球根贸易而导致哄抬价格的悲剧再次发生，由利瑟小镇郁金香种植者联名倡导，把球根的生产和贸易纳入了专业的框架下，于1860年由种植者联盟，在利瑟成立荷兰皇家球根种植者协会（KAVB）。球根的贸易是以签订书面合同的形式来销售的，合同则是按照球根行业内规定的贸易协议来签订。KAVB还设有处理专业种球销售商和购买商之间的仲裁委员会，就像法庭一样，双方都有机会将各自的不同意见和分歧提交给行业仲裁庭进行仲裁。随着郁金香等球根贸易的扩大，协会又成立了荷兰批发商协会，建立了专业的球根经纪人、球根拍卖市场机制，不仅在种植者和经销商的交易中扮演着重要角色，还负责交易活动的财务结算，确保种植者和经销商之间的贸易既专业也诚信。球根经纪人会根据不同的品种确定相对稳定的市场价格，每年球根种植者协会和荷兰批发商协会，要对当年的球根收成好坏等情况进行评估，评估收成结果一定要事先提醒买方，让买方能尽早预估来年郁金香等球根价格变动的幅度，保障贸易的价格诚信和支付诚信。

市场上各品种郁金香的球根
The variety of tulips in the market

利瑟小镇街道
The streets of Lisse

无处不在的郁金香图案
Ubiquitous tulip pattern

无处不在的郁金香图案
Ubiquitous tulip pattern

利瑟小镇街道
The streets of Lisse

利瑟 LISSE　　　　　　　　　　　PRODUCT TOWNS

利瑟人的守信境界已经成为他们引以为傲的一面。郁金香可任由人们意愿，安排盛开时间。利瑟小镇的种植者和科研人员让郁金香的守信达到了极致。他们可以通过对郁金香的精心栽培、储藏和促成时间上的控制与处理，让郁金香等绝大多数球根花卉，在任何一个确定的时间点准确绽放花朵。具体的说：10月17日是你的生日，你希望在这一天能有郁金香为你绽放，你把这个愿望告诉利瑟小镇的种球商，他们给你一些订制处理过的郁金香种球。只要按要求种植的时间和温度种植，到你生日的那一天，你会惊奇的发现郁金香准确地为你盛开。神奇程度让人赞叹，至今没有任何科技可以让其他花卉与之媲美。这种梦幻般的技艺，专业上称"促成栽培"，民间则称作是"上帝以神物对人类的信守"。世界各地的园艺学者无不对此技艺惊叹和羡慕。笔者曾经陪同国内的花卉专家参观利瑟小镇的球根生产商，当提及讨教学习促成栽培技术时，接待者风趣地说：这不是技术，是球根种植者心灵与郁金香对话的结果，它是利瑟人内在守信境界的物像反应。

让荷兰国花绽放的更加艳丽灿烂

第一次世界大战以后的10年，郁金香等种球贸易得到了前所未有的发展，此次成功的关键驱动力是国际范围的广泛宣传。宣传不仅范围广，而且十分专业。在全世界的著名公园、展览以及各种园艺联合会，都能见到荷兰的郁金香。从那时起，就奠定了郁金香作为荷兰象征的基础。如今，郁金香几乎成了荷兰的代名词。全世界各地纷纷组织以郁金香为主题的文化交流活动，可以说没有一个荷兰主题的活动能缺少郁金香的，真正做到了"无郁金香不荷兰"。郁金香作为荷兰国家的象征，牢牢扎根在全世界人民心中。以郁金香为代表的荷兰种球所带来的文化功能不言而喻。而那个从郁金香引种开始，为它坚守近400年的利瑟小镇无疑成为郁金香---这个荷兰国花最忠实的代言人。

无论是乘车还是骑车去利瑟小镇，都会在N208的主路与一条支路交叉中央，看到一个不大的转盘，转盘中央花坛上是一组单体球茎剪影图案组合成郁金香花朵的雕塑，设计巧妙而简洁。交叉口支路就是通往利瑟小镇的入口。

利瑟小镇处处充满着郁金香为代表的球根花卉意象。小镇的街道，有许多是以花卉名称命名的；小镇的街景标识，建筑立面甚至地面以及建筑维修维护的挡板都呈现出球根花卉的图案和装饰。漫步小镇中心，无论街边绿地、路旁绿化，还是沿街民居花园，都开满了五颜六色的球根花卉；穿梭于小镇的各个商店，各色商品琳琅满目，大到油画、地毯、家具、墙纸等，小到花瓶、瓷砖、银器、木鞋等，再到包袋、领带、雨伞、邮票，都可以体验到郁金香无处不在。

库肯霍夫公园
Keukenhof park

库肯霍夫公园
Keukenhof park

库肯霍夫公园中的白孔雀
White peacock in Keukenhof park

库肯霍夫公园
Keukenhof park

利瑟 LISSE   PRODUCT TOWNS

第二次世界大战以后，花卉旅游业逐渐繁荣起来。利瑟小镇从1947年起开始在每年4月最后一个周六举办花车巡游，以这种方式感恩花神带给小镇人们的吉祥与富裕。当时小镇的人们还希望为郁金香建造一个属于神物归属的极乐花园，也为了向世界展示郁金香——这个自然界最迷人的创作奇迹。1949年在利瑟镇长的倡议下，以1830年德籍景观园艺师设计的库肯霍夫公园为基础，将库肯霍夫公园改建成了一个开放式的室外球根花卉展示基地。

小镇的规划布局紧紧围绕球根花卉的生产贸易功能合理地安排着：中心由两条主要街道划十字组成，中心以外的街道错落有致地布置着小镇居民的房子和郁金香等球根植物的生产和贸易公司，中间则是运河河道，外层连接了球根生产公司加工和储藏球根的巨大仓库。整个小镇的外围被五彩缤纷的地毯般的球根生产花田、花海所包围。花田的核心位置，就坐落着让全世界迷恋的郁金香展示公园——库肯霍夫公园。

公园设计布局以英式风格为主，有7个具有不同特色的主题园。如：灵感花园由大小各异的花园组成。每座均拥有自身的特点和主题，从古典到超现代感，风格百变。艺术公园则是由荷兰著名艺术家创作的数百座雕塑及艺术品点缀。公园自然优雅的环境巧妙地映衬出了这些艺术品的美妙绝伦。公园内活动丰富多彩、特色鲜明。在这里，人们可以看花赏鸟。

园内郁金香的品种、数量、质量以及布置手法堪称世界之最。各类花卉达600万株以上，还有很多难得一见的珍稀品种，各种树木2500多棵，种类达87种。园内还放养了30余种鸟类和小动物，水池中有天鹅在悠然游弋，草坪上有孔雀在轻盈漫步，树丛间有松鼠在欢快跳跃。除此之外，主题花车巡游、每日花展、游乐摄影等活动也丰富了游客的体验。

利瑟花车巡游
Flower Parade Lisse

游人如织的库肯霍夫公园
Keukenhof park is filled with visitors

花田骑行
Cycle in the flower field

鲜花马赛克展览
Flowers mosaic exhibition

花田泛舟
Boat among the flower field

利瑟 LISSE PRODUCT TOWNS

自2006年起，库肯霍夫公园开启了一个新的传统，每年公园都会选择一个特别的主题，如：2008年主题是中国奥运会，2013年主题则是英国的英式花园等。这种变化可以更加丰富地展示花卉的种类，花卉每年都会以特色新颖的形象示人，这无疑有助于聚集客流，使公园每年保持大量的参观游客。此外，五月的第一个周六都会举办"鲜花马赛克"展览和比赛。鲜花马赛克作品多达百余个，每一个画作至少要用万朵花瓣组成，各式各样美丽而有趣的鲜花马赛克令人目不暇接。

花田骑行也是十分受欢迎的项目，游客只需在公园门口租一辆自行车便可以开始花田骑行。骑行过程中可以参观球茎生长的黑郁金香博物馆，还可以随时停车拍照，将自己的身影留在郁金香花田这块唯美的画布上。花田泛舟则是乘坐"私语小舟"深入花田深处，从另一个视角欣赏花海。百年风车观景台也是不错的选择，登上风车极目远望，斑斓的色彩充斥着眼帘。

如今，利瑟小镇的郁金香公园已总计接待超过6000万来自世界各地的游客。小镇的郁金香等球根的生产总值达5.4亿欧元，总的出口额5亿欧元，出口世界100多个国家和地区。荷兰利瑟小镇——一个以郁金香为特色的小镇，正在用艳丽灿烂的荷兰国花，装点着人类的生活环境，用他们的坚持与守信，维护着荷兰国花所映射出的荷兰人诚信的精神境界。

编者感言

利瑟的故事是让人感动的，这是一个颇为治愈的故事。其实家中的阳台也有一盆来自荷兰的花卉，这是一盆青蓝之色的绣球花。购买之时价格不菲，但在花丛中一眼就相中了它，花盆外围的一张围绕的贴纸上面是荷兰花农的悉心打理植物的形象。如今，许多花都枯萎逝去，但这盆绣球花却依然绿意盎然。哪怕一段日子无人照料，它也依然坚持着。

花是有情而通人性的。有时甚至觉得，花比人更多情。花多少时间在它身上，它都会一一回报你。也许比之观赏花开，等待花开才是最幸福的时光。正是由于利瑟人们对花的坚守，花神才会给与相应的回馈。在哪里跌倒，就在哪里爬起；失于诚信，那就立于诚信。利瑟的勇气不是所有人能具备的。花农用心与花对话，精心地照料。也只有这种坚韧和坚持的花匠才能重新开始，依然如初。

感情不可存银行。感情应该像养花一般，需要精心观察，施肥浇水除虫，随机应变，才能绽放出美丽的花朵。希望我们的梦想和感情都能绚烂开放。

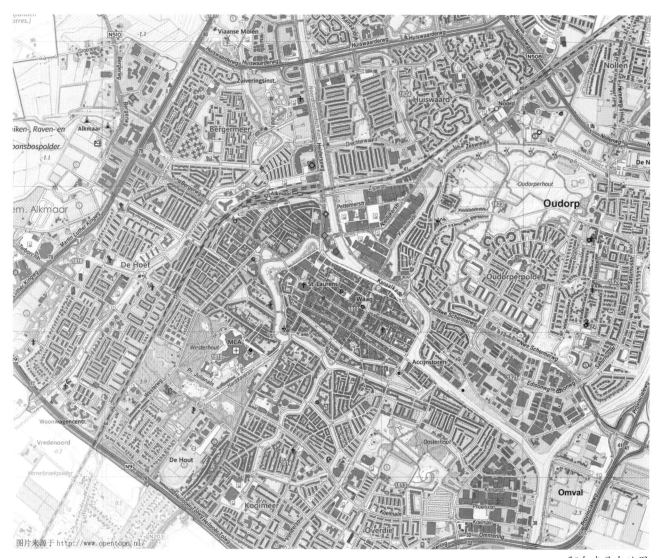

阿尔克马尔地图
Map of Alkmaar

阿尔克马尔 ALKMAAR　　PRODUCT TOWNS

阿尔克马尔街道和河道
Alkmaar street and river

一座诉说着胜利的小镇——阿尔克马尔

Victory，是给一座城镇披上五彩云袍般的美誉。

胜利，首先来自好运。那场战役，真的有太多的天时地利人和，太多的刚刚好。

但胜利，绝对不能只依靠好运。阿尔克马尔的胜利其实写成一部小说乃至拍成一部电影都可说是绝对的精彩绝伦。两军交战，只靠勇武和人多显然是不够。阿尔克马尔的胜利可谓是智勇双全的典范，不但有阿尔克马尔人民的骁勇善战，也有外围城镇的妙计联盟。总之，若遇上罗贯中的妙笔描写，不会比《三国演义》里的赤壁大战逊色多少。

从此，幸运女神与胜利女神联袂相伴，不断眷顾着这座城市，智慧、勇敢、坚毅的阿尔克马尔人民书写了一篇又一篇历史华章。

在荷兰民间人们常说,"胜利始于阿尔克马尔",这句谚语源于在荷兰抵抗西班牙的80年战争中,那是发生在阿克玛尔小镇的一段英勇的故事。

1572年荷兰人因为对西班牙国王高税赋、高集权的不满,以及反抗西班牙对新教的残酷镇压,由民间起义开始很快爆发了战争。1573年8月,为了镇压荷兰的起义与反抗,由费尔南多·阿尔瓦雷斯·德托莱多率领的数以千计的西班牙士兵,连续攻占了几个城市后包围了阿尔克马尔。勇敢的阿尔克马尔人民将滚烫的焦油和燃烧的树枝从城墙上扔向敌人,这是阿尔克马尔历史上最荣耀的时刻之一。西班牙军队希望通过断绝阿尔克马尔粮食供应的方式困死城内的起义者。一位勇敢的木匠成功地将求救信送出城外,最终外援地区通过开闸放水的方式,使阿尔克马尔周边圩田的水位提高,西班牙士兵纷纷落荒而逃。这就是"胜利始于阿尔克马尔"的来历,也是荷兰威廉王子领导起义者第一次战胜了西班牙军队,阿尔克马尔也因此成为荷兰第一座脱离西班牙统治的城市。围攻结束后的10月8日,成为阿尔克玛尔的庆祝日。此后,荷兰人民继续抵抗西班牙,历经了80年战争,最终实现了国家独立。

阿尔克马尔是一个具有辉煌历史和丰富文化的城镇。小镇内河道纵横,一条运河从小镇的中心穿越而过,阳光让倒映于河畔的中世纪建筑,闪着点点金光,微风拂过河面,泛起粼粼细浪,古镇阿尔克马尔呈现着辉煌与古老的魅力。阿尔克马尔的黄金时代要从镇中的大教堂说起:圣劳伦斯(St.Laurenskerk)教堂,由来自梅赫伦的建筑师安东尼斯·凯尔德曼(Anthonius Kelderman)在1440～1512年间设计建造。教堂为布拉班特的哥特式风格,采用26个立柱,来支撑建筑内的拱顶和天花板。严谨的结构与教堂华丽外表以及精雕细琢互相辉映。教堂装修富丽堂皇,但也颇为庄严肃穆。这个教堂是阿尔克马尔古典建筑的代表和典范,也是阿尔克马尔黄金时代繁荣昌盛的标志。

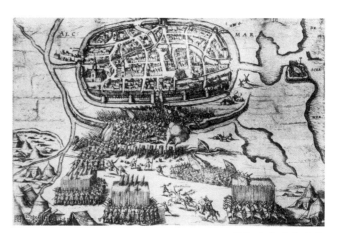

1573年阿尔克马尔战争
Battle of Alkmaar 1573

阿尔克马尔 ALKMAAR

PRODUCT TOWNS

十世纪时由渔村发展为集镇的阿尔克马尔，在1254年就已经获得立城许可。几个世纪以来，贸易和水路转运集散，让阿尔克马尔人民进入了富裕、繁荣的黄金时代。1630年后，随着北荷兰地区几条大湖的干枯，阿尔克马尔失去了连接南海的便利通道，海上贸易逐渐减少，小镇周边却增加了大片圩田，圩田地区土地潮湿、多雨多雾，加之小镇本来就气候温凉、日照偏少，使得多汁牧草的生长更为有利。原来依赖集散贸易的阿尔克马尔，不得不向畜牧业的产业发展方向转移。多代的贸易经验让阿尔克马尔人清楚地意识到，市场是产业发展的关键。在全面开展畜牧业生产的同时，努力向国家争取到了乳制品的交易权利，终于在1365年阿尔克马尔镇获得了颇具权威性的度量衡工具——乳酪过秤权利，随后在小镇中心建起了奶酪交易市场。在日后的发展中，阿尔克马尔人再次用他们的商业智慧和努力，将奶酪交易市场培育成了荷兰最大的奶酪交易市场，从而稳固了阿尔克马尔奶酪交易的领头地位，为小镇赢得了"奶酪之镇"的美誉，也为阿尔克马尔镇持续的繁荣与发展产生了深远影响。

阿尔克马尔小镇的人们有着崇尚自由、骁勇抗争的品格和文化内涵。阿尔克马尔是在荷兰率先倡导信仰自由，实现多教共存的宗教信仰新制度的小镇。1520年前，阿尔克马尔小镇的人们都信奉天主教。在圣劳伦斯教堂建成后不久，有越来越多的人希望改革天主教，建立天主教新教。在威廉王子的带领下，主张宗教信仰自由的人们与天主教徒之间，发生了宗教之争，新教徒占领了阿尔克马尔的大教堂和修道院，为新教徒建立了新教规和新仪式。从此，荷兰信仰的"一城一教"的宗教信仰制度被打破，多教共存的宗教信仰新体制由阿尔克马尔小镇的人们开启。

另外值得一提的是，一段阿尔克马尔人在卡斯特里克姆战役中打败英俄远征军的故事，被刻在了法国巴黎凯旋门（Arc de Triomphe）上，让世人永远记住这个骁勇善战的小镇人民。

阿尔克马尔圣劳伦斯教堂
Church (Sint Laurenskerk)

古运河
Ancient canal

古运河
Ancient canal

古运河
Ancient canal

市政厅塔
Townhall tower

阿尔克马尔 ALKMAAR PRODUCT TOWNS

漫步于"奶酪城市"的街间,除了享受奶酪店飘来的浓郁香气外,不时出现的各色博物馆、展览馆,也让小镇充满着文化的气息。阿尔克马尔镇内有400多处国家级文物点,镇中古朴典雅、精致壮丽的中世纪建筑与纵横交错的古运河相得益彰,景色分外迷人。如今,最吸引游客的莫过于最具传统风格的奶酪交易会表演:表演会于4月至9月每周五上午10~12点举行,固执地沿袭着有三四百年历史的古老奶酪交易,没有机械、没有汽车、没有电脑、传真机和手机,甚至没有人使用计算器,奶酪交易会的表演带着游人,穿越到小镇400年前的时代里。

阿尔克马尔奶酪市场设在waagplein(过秤广场),上午7点奶酪摆放者开始工作,约3万公斤金灿灿的圆形奶酪,在过秤广场整齐排列,每次大约有2400个奶酪。如果遇上下雨或者温度超过28摄氏度,在奶酪下面会垫上一层特殊的帆布。奶酪搬运工必须在9:30之前到达,迟到、犯规者将会被公示及被罚款,罚款收入一部分用来资助苏里南同名城市阿克玛的一所学校,其余部分作为奶酪搬运工工会收入。每个奶酪搬运工身着白衣白裤,头戴四种颜色的草帽:红色、蓝色、绿色和黄色,与船型奶酪木架的颜色一起,分辨出主导奶酪市场的四个仓储公司。唯一带着橙色草帽的是四个仓储公司的市场代理人,被称为"奶酪之父"。市场开始前"奶酪之父"会告知搬运工当天的奶酪交易量、当天的重要客人,分配当天每个仓储公司负责搬运的范围。

上午10点摇铃开档,市场上钟声响起,标志着奶酪市场正式开始。摇铃开档仪式由小镇邀请影星、著名运动员、外国使节等名人,或重要生意伙伴主持,小镇上星光璀璨,也因此吸引了很多好奇的游客。

Waagplein 广场
Waagplein square

奶酪搬运工戴着不同颜色帽子
The cheese porter wear different colored hats

奶酪搬运工戴着不同颜色帽子
The cheese porter wear different colored hats

广场上交易的奶酪
The cheese is traded on the square

广场上交易的奶酪
The cheese is traded on the square

广场上穿着传统服装的姑娘们
The girls wearing traditional clothes on the square

阿尔克马尔 ALKMAAR　　PRODUCT TOWNS

通过看、闻、敲、切、钻、捻、尝等方式来评判奶酪的味道、脂肪量以及软硬度，也可切开奶酪来看里面的孔洞即"眼"的数量和均匀程度，"眼"是奶酪发酵过程中产生的有益乳酸菌引起的。完美奶酪里面的眼应该是均匀的。督察员鉴定奶酪质量后，各家的奶酪搬运工将奶酪从过秤广场运到过秤房，在过秤房检验奶酪的品质，买卖双方以不断地拍击对方手掌来讨价还价，交易完成也以击掌为定，不需再签订契约。随后奶酪便送到过磅秤上称重，打上标志。交易成功后，奶酪搬运工成为全场的主角，两人一前一后，通过皮质吊带挑起放置8个圆形奶酪船型担子，摆动双手，矫健熟练的步伐穿梭在过秤房和广场之间，那夸张逗趣的动作正是游客们最爱的一道风景。

广场中的另一大亮点是那些传统服饰的俏丽荷兰美女：白色风帽，红色围巾，蓝色围裙，脚蹬木鞋，挎着竹篮，里面放着一袋袋小块奶酪，向周围的游客人群兜售着。人们除了观看生动的奶酪交易，更少不了品尝一番当地的特色奶酪。奶酪交易结束后，便会运往全国乃至世界各地，将奶酪集市上空弥漫着的香味，带给远方的人们。

每年的阿尔克马尔奶酪市场都会吸引30万的观光客。如今，徜徉在阿尔克马尔中心，倾听一座座建筑诉说着阿尔克马尔厚重的历史，不禁感慨：胜利依然始于阿尔克马尔！

编者感言

阿尔克马尔是很会反抗的城镇，对西班牙战争的首胜也好，打破一城一教的先例也好，被雕刻在凯旋门上也好。这无疑是一个颇为善战的城镇。但如今，它最出名的竟然是奶酪。这总让人觉得颇为不符。奶酪总是柔软温暖的存在，入口即化的那种温存和浓郁的奶香总能通过味蕾直入心房。

可以发现阿尔克马尔不是那么刀光剑影、铁马冰河，是芝士的温柔乡。阿尔克马尔的人们其实很纯粹，也特别真实。当他们欢快地用着古法在广场上进行着奶酪交易时，也许他们只想拥有一个自由美丽、真正属于自己的城镇。

幸福有时候是需要去争取的。也许只要再努力一点点，两位女神就会同时来到你的身边。想起一种关于"V"字手势起源的传说[1]，不得不感叹，有的时候温柔故乡也需要用自己的弓箭去守护。

[1] 代表胜利的"V"字手势源于英法百年战争，法国人扬言砍掉英国弓箭手的中指和食指，让他们不能再张弓射箭。但后来英军打败了法军，英国人便伸直中指和食指示威。

荷兰案例 Cases from the Netherlands

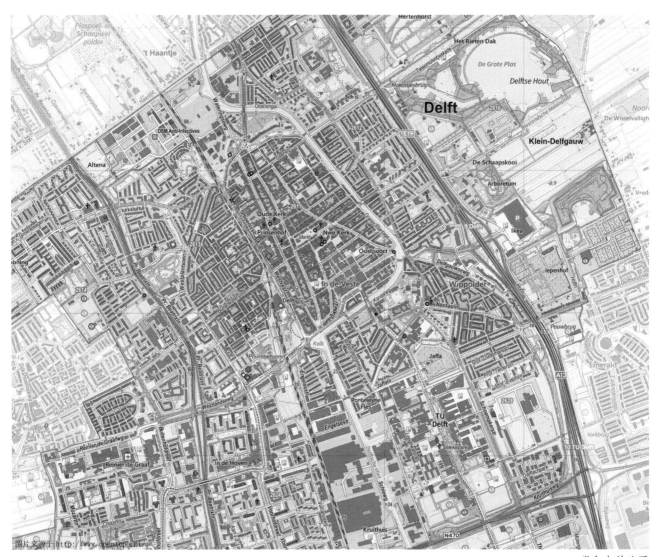

代尔夫特地图
Map of Delft

代尔夫特 DELFT　　PRODUCT TOWNS

代尔夫特大市场
Grote Market Delft

蓝天下的瓷器（china）——代尔夫特

但凡去过欧洲就会对那个地区的城镇布局深有感触，欧洲许多小镇总是以教堂为中心，然后慢慢发展起来的。有教堂的地方就有落脚的地方。代尔夫特也是如此，由教堂开始发展起来，整个城市被沟渠围绕。这个城镇的教堂不但带来了城市的发展，也是荷兰皇室的安息地。所以，对于整个荷兰来说这座教堂也是举足轻重的。

不过，其实让世界许多人熟识这座小镇的其实是蓝瓷和那幅让人特别喜欢的画——《戴珍珠耳环的少女》。

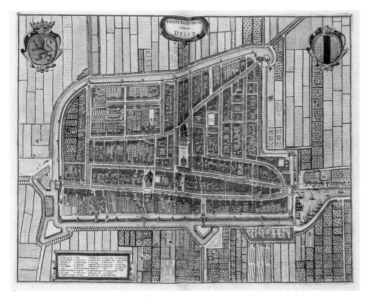

代尔夫特新教堂
Nieuwe kerk Delft

1650年的代尔夫特：荷兰共和国中部强大的城市
Delft 1650: Powerfull city in the middle of the Dutch Republic

1900年代尔夫特乌斯特普尔特：东部城市入口
Delft Oostpoort 1900: Eastern entrance of the City

代尔夫特老长堤
Old Long dike Delft

代尔夫特 DELFT

PRODUCT TOWNS

地处海牙和鹿特丹之间的代尔夫特是荷兰小城中最优雅和最具特色的典范。小城的形成与荷兰大部分的古城一样，是以教堂为标志物和中心点，然后向外慢慢扩延。古城中心是由新旧两个教堂和市政厅等建筑围合而成的四方形广场。广场散发的一种古朴气息，与四方形态一起让人产生一种庄重和威严感，似乎是在提示着这座古城镇与皇室之间的联系。

在西班牙王朝统治时期，荷兰共和国第一执政王威廉·奥兰治带领尼德兰人，与西班牙王朝统治者整整抗争了几十年，最后这位执政王被刺死在了古城的王储园中，后来代尔夫特市中心的新教堂被选为这位执政王的安息之地，此后王室成员死后遗体都照例安葬于此新教堂。古镇起源于一条名叫"Oude Delft"的人工水道，荷兰语"deft"的动词"delven"其实是挖掘的意思，名从其意，代尔夫特意指古代人以挖掘水道，筑起堤防挡住北海，并用水道来排干沿海的沼地，形成由水道环城护城的古镇。这个名字充分反应了居住于这个古镇的人们长期以来与外敌和自然抗争，形成了不屈不饶的精神品格。

不过代尔夫特也是颇具文艺气息的小镇，画家维美尔一生居住和作画于此地。"傍晚的斜阳温暖地照耀着小镇，泛红的晚霞随意涂抹着天空，在波光粼粼的水道旁，代尔夫特进入一天中最美的时光"。就像荷兰维米尔（Vermeer）笔下那幅《代尔夫特城风光》一样："蜿蜒的水道、白色的吊桥、绿色的橡树，在夕阳下呈现着无比美丽的光影"。城镇随处可见的画廊和艺术店铺，与古镇中心定期举行各种"音乐节""戏剧节"等文化活动一起，叙述着小镇悠远精彩的历史、艺术与文化。

代尔夫特风光，维梅尔，1661年
View on Delft : Painting by vermeer, 1661

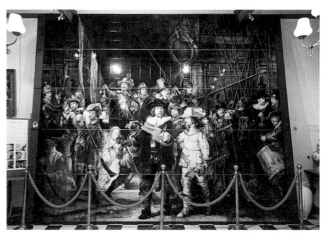

《戴珍珠耳环的少女》蓝瓷画
"Girl with a Pearl" Blue Porcelain Earring

《夜巡》蓝瓷画
"The night watch" Blue Porcelain Earring

代尔夫特蓝瓷
Delft Blue Porcelain

代尔夫特随处可见的蓝瓷商店
Delft Blue Porcelain Store in Delft

代尔夫特 DELFT　　PRODUCT TOWNS

特色产业

荷兰一直由东印度公司向中国采购青花瓷工艺物件，却因中国17世纪后期内战频繁而中断。精明的荷兰人开始研究仿制中国的陶瓷，在完成了解密钴与锰的配方后，青花瓷的神秘迎刃而解。然而执着追求的荷兰人并没有因此而满足。他们又从日本的彩釉技法中学到了核心，把它们与中国的陶瓷技艺巧妙地融合，并形成了举世闻名的国宝产业和精美的艺术产品——代尔夫特蓝瓷（Delft Blue）。

探讨关于这个例子对中国特色小镇的借鉴与参考，笔者无心于蓝瓷的仿制过程，而是分外关注外来产业如何能在本土发扬光大，并在几个世纪中经久不衰的根源。显然地方文化的注入给外来产业赋了新的内在灵魂，以至于这种新生的文化可以在当地的土壤中能扎根、生长。

蓝瓷的基调文化根基是"蓝色"。生长在北海边的荷兰人，在长期与大海抗争过程中，也深深依靠着海洋，蓝色已深深地刻在他们民族的骨髓里，可以说荷兰是崇尚蓝色的国家，只是有时被靓丽的橙色郁金香的激情深藏了起来。蓝瓷的产品形式都来自荷兰的生活，诸如风车、木鞋与郁金香等荷兰典型的"标志"造型是蓝瓷中的主角。蓝瓷上丰富多彩的图案更是荷兰人700年文化生活的影像和艺术创造的缩影。可以看到，一个能反映民族灵魂、刻有生活印记、体现标志韵味的产品，成为国宝产品而蜚声国际是不足为怪的！

此外，蓝瓷的创新精神也值得肯定，如为了解决伦勃朗《夜巡》原作的绝妙用光的效果，工匠们钻研了许久。从调整颜料配置比例，烧制不同深浅层次色彩，在蓝色的基础上变幻出各种深浅颜色等，前后共用了一年多时间才解决了所有的难题，最后用480块烧制的瓷片拼制完成这幅名作。蓝瓷是可以订制生产的，包括皇家订制，也会以"国礼"的身份赠与其他国家元首，这大大提升了品牌形象和档次。蓝瓷也可由游客以自己的想象，设计创作出属于自己独一无二的作品，既体验了参与的乐趣，也增加了产品的销量。这些在中国特色小镇的产业建设和发展中都值得借鉴。

代尔夫特蓝瓷
Delft Blue Porcelain

代尔夫特街景
Delft Street View

代尔夫特市政厅
Delft Townhall

代尔夫特运河
Delft Voldersgracht

代尔夫特理工大学礼堂：学生聚会场地
Aula TU Delft: Gathering venue for students

代尔夫特 DELFT　　　　PRODUCT TOWNS

小镇古朴典雅，充满活力

清晨薄雾中的小镇，恬静迷人。许多野生的天鹅、海鸥、野鸭、大雁等在古城遍布的河流、水道中自由优雅地游动和嬉戏。坐在古老的运河边，看着白天鹅和野鸭子等着人们喂以食物，一幅人与自然和谐的画面，让人感动；代尔夫特生机盎然，还与代尔夫特理工大学与古城完全融合共生有关。代尔夫特大学世界综合排名第65位，建筑、规划与环境专业一直在世界排名前五位，堪称世界名校。整个校园呈开放式，教学楼、院系布局分散，大学俨然成为小城空间的一部分。城镇街道、商业等公共设施，既为大学也为城镇居民提供服务。城镇和学校就这样有机地联系在一起。在校大学生与市民们共同生活，让知识与经验、历史与现代可以产生更多的交流与碰撞，融合成新的文化在这座古老的城镇轻松而美好的发生、发展和传递下去。

古镇内街，道路尺度宜人（宽约5～8米），两旁点缀着中世纪及文艺复兴时期典型的建筑，沿古城运河以同一种风格纵横展开，棕色为底，白、橙相间，每个建筑细节各有差异，却也精雕细琢。博物馆、纪念馆与古董、蓝瓷商店穿插其中，温馨宜人，韵味十足。小镇的中间道路是供行人和骑车赏景的领地，偶尔有汽车闯入，使人有在古代、现代之间穿梭的感觉，这也是荷兰独创的"乌纳夫"交通模式。让户外逗留和交通、街道综合起来，使小镇交通、安全和趣味相结合。

代尔夫特小城现在以代尔夫特大学的EWI电子大厦为新城的至高点，另以大学礼堂Aula、中央图书馆和城市新火车站的建筑形成新地标。由荷兰建筑师雅各布·巴克玛（Jacob Bakema）设计的Aula建筑，因为飞碟状的巨大悬挑形式使它具有了"UFO"的外号，也被认为是粗犷主义混凝土建筑风格的代表。由荷兰建筑事务所Mecanoo设计的中央图书馆和代尔夫特的新火车站也是不错的经典案例。前者以其大面积的生态草坡屋顶体现了人与自然共生的理念而闻名。完成设计不久的新火车站，其架在混凝土高架桥建筑外层的玻璃层可以即时地反射出小城的天空，仿佛是流动的画布巧妙地装饰了原本单调的钢筋水泥结构，透过窗口也可以纵览整个古城。这种富有想象力的科学设计让新城古镇完美结合了起来，成为古城改造的典范。

图片来源于 http://www.mecanoo.nl

代尔夫特地下火车站：通往老城的新入口
Undergrond trainstation Delft:New Acces to the old Town

代尔夫特理工大学：世界知名理工大学之一
TU Delft Campus: One of worlds renown technical Universities

代尔夫特图书馆：校园核心
Library Delft: Heart of the Campus

代尔夫特图书馆：大生态坡屋顶
Library Delft: Large ecological slope roof

代尔夫特 DELFT　　　　　　　　PRODUCT TOWNS

代尔夫特大学图书馆始建于1997年，由荷兰Mecanoo建筑事务所设计完成。这个图书馆造型非常别致，以其巨大的生态草坡屋顶而闻名。图书馆其它几面皆为巨大的玻璃幕墙，顶面则是一个大草坪，斜延伸直至与地面合为一体，成为一道别具一格的校园风景。建筑师显然希望学生们充分利用这块草地，在充足的阳光和绿地中学习并获取灵感。这个设计中非常值得一提的是一个高达40米的圆锥体矗立在屋顶上，白天可以为图书馆内提供充足的光线，晚上则会像一座灯塔散发光芒。代大图书馆因其特别的造型和设计，每年吸引着大批学生和游客围观。这样的校园建筑也算没有辜负代尔夫特大学在建筑学的领先地位！

新地标建筑，是小镇从历史走向现代化过程的写照，因当今需要而自然产生，却能完美地携带着古镇的固有基因，注入了古镇居民几百年来形成的粗犷、挑战、创新的精神内涵。这种因需而生的建筑发展模式是值得中国在小镇发展中深思和参考的。

编者感言

说到瓷器，不难想到景德镇。瓷器(china)对于中国人来说蕴含的情绪实在太多了。代尔夫特因为崇尚中国的瓷器，所以开始了钻研，而今也有了自己的一片天地。所以引进外来的文化元素并不是不好，只是要如何让他们在自身的土地上生根发芽，这需要深谙两种文化的人来融合，接了地气后，才能茁壮成长。代尔夫特还是如油画中走来一般，虽然他也有很多现代的经典建筑，不过点缀在这里一点也不让人觉得突兀。而是觉得他们就该如此，本就属于这里。

现在的城镇建设中经常会遇到一个问题。我们的设计规划经常容易变成要不全部拆掉重来，要不全部复古还原。那么在这种来来回回全盘往复的过程中，其实有许多资源被浪费，拆掉的自然是不能再恢复过来了，那么就要去别的地方收旧来补充，收来的自然也是别的城镇拆掉的。这种拆旧补旧的方法是否真的可行呢？其实只是"拆东墙补西墙"，结果还不是破坏和残缺么？

如果有一天，旧石板用光了呢？旧的所谓的不重要的全部被拆光了呢？现代和古代相结合的难道不可以么？发展所谓的CBD就一定要拆迁所有的地区房屋么？其实东墙西墙就是在这样没有可持续性的发展和规划下被双双破坏的。古今结合，和谐共存，其实也是一条值得借鉴和探索的道路。

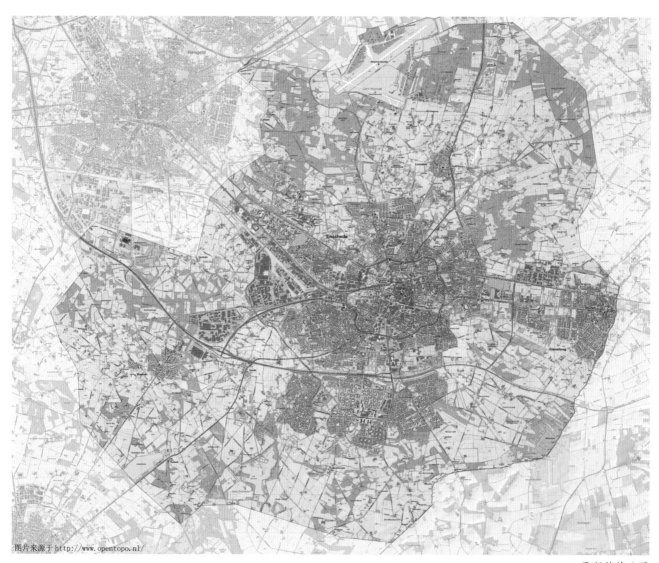

恩斯赫德地图
Map of Enschede

恩斯赫德 ENSCHEDE　　PRODUCT TOWNS

恩斯赫德经典建筑
Classic architecture in Enschede

不畏切磋成美玉，不断琢磨出新品——恩斯赫德

　　在看阿尔梅勒资料的时候，就特别喜欢这种没有开发商参与的规划设计模式，原来它是师承恩斯赫德。命运的旋转木马，总是有些无常。对于恩斯赫德来说，灾难也许让世人多少对它有些同情的气息。这倒是让我想到了《你的名字》，电影中那个城镇因为流星就此陨灭。恩斯赫德经历的苦难也可谓深重，重建也是直到2010年才全部完成。但城镇毕竟是人们往日的家园，现实生活中，人们总会继续复建和重修，毕竟这是自己的故土。也正是这种生生不息的执着，让恩斯赫德重新建造起来了，也开发了一种让居民更加融入其中的规划建造方式。参与其中，爱和责任感势必将更深入肌理，乃至心肺。

　　命运的旋转木马，看似随意，但是一人、一城，总不会那么完美无缺。接受命运的圆舞曲，所谓的天降大任于斯人也。潇洒转身，如切如磋，如琢如磨，才成就了这块绝世美玉。

恩斯赫德小镇内的工厂
The factory in Enschede

恩斯赫德城镇中心
The city centre of Enschede

特温特大学
University of Twente

恩斯赫德 ENSCHEDE PRODUCT TOWNS

一个城市如何能够度过一次、甚至两次具有灾难性的打击？荷兰东部的恩斯赫德小镇通过弹性设计的建设实践，回答了这个问题。小镇的第一次困境，是在将大量的工厂外包给劳动力成本较低的国家后，整个经济都被改变了。第二次是烟花仓库的爆炸，使得小镇中心生活区的北部炸出了一个63英亩的大坑。恩斯赫德不仅度过了这样的打击，还在实现经济多样化、居民参与式的灾后重建方面，树立了良好的榜样。

恩斯赫德不是一个典型的荷兰小镇，因为它没有河流或防御工事。直至19世纪后期，恩斯赫德经过工业化变成了纺织和啤酒生产的中心。小镇是在区域道路和早期铁路线相交的地方逐渐发展起来的，这条铁路线将阿姆斯特丹和鹿特丹与欧洲内陆连接了起来。当地产业的鼎盛期出现在1930年左右。市中心的许多地区是在这一时期形成的。在历史悠久的中心地区周围，美丽的1930辛格路（Singel）(Singel是一条环形路，建于1930年，有着良好的景观和树木，旨在更好地接入进入城市的第一批车)围绕着城镇，成为了解和体验这座小镇的好地方。

恩斯赫德的工业发展使其成为早期荷兰的工业中心。通过一张著名的历史图片可以看到：大型工厂、耸立的烟囱群、村舍农田等元素有机融合在一个空间中，构建了其独特的乡村景观结构。同时，恩斯赫德还有一些花园村庄形式的工人住房。其中，令人印象深刻、广受欢迎的有帕特莫斯（Patmos）和德莱瑞思（De Laares）。恩斯赫德的纺织大亨在工人住宅和美丽城市公园的建设中做出了巨大贡献。

但是从1960年开始，大量的工厂倒闭了，这使大约3万名员工在20年的时间里相继失业，导致了严重的社会危机和发展困境。许多工厂被拆除、改建或直接关闭。然而，与此同时，从60年代开始，政府在恩斯赫德创建了大学，后来变成了科技学院，也就是如今的特温特大学。

大学技术的商业化应用，为恩斯赫德提供了一个新的经济视角。如今，大学和其附属公司及机构就像恩斯赫德最大的雇主，吸引了荷兰北部地区的许多工人。新的足球场建成后，恩斯赫德似乎重新找到了出路，为迎接21世纪做准备。然而，磨难并没有就此停歇。

恩斯赫德工人住宅
Enschede workers residences

2000年恩斯赫德烟花爆炸灾难
Day of the fireworks explosion Bruijn, 2000

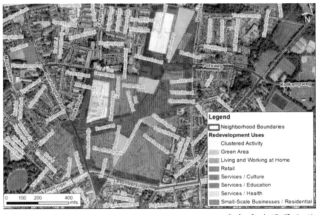

城市重建愿景文件
Vision document

旧建筑和新建筑的融合
Integration of old and new buildings

自建房屋
Self built Houses

恩斯赫德 ENSCHEDE   PRODUCT TOWNS

2000年5月13日，灾难再次来袭。一个烟花工厂和储存设施爆炸了。爆炸的工厂位于老工业区和住宅的混合区，多人死亡，1000多人受伤，在市中心北部炸出了63英亩的深坑。这是战后荷兰发生的最大的灾难之一。访问灾区后，荷兰当时的总理维姆•库克承诺"没有人会被落下！政府将聚集资金和力量，用于灾害重建"。

爆炸发生后的一个主要问题是：谁负责城市重建的协调和设计工作。到2000年底，伟大的荷兰建筑师德布鲁因(Pi de Bruijn)负责重建工作。德布鲁因是地道的恩斯赫德人，同国民议会的设计师及阿姆斯特丹大型城市项目的负责人一样著名。德布鲁因作为政府和商业之间的协调人，凭借其影响力号召市民也参与到了规划过程中。

虽然很多人失去了家园，但仍然拥有土地。居民们对那些被大型开发商冷落的地块进行了独特的设想，决定由自己来建造房子。因为根据荷兰的传统，城市发展是一个联动的整体行为。由此，诞生了一种独特的设计建造方法。而此地的近1800栋房子都是通过这种方式建设而成的，大家从三个层面参与到规划设计过程中。

首先，设计之前，设计师对所有居民和利益相关者进行访谈，询问他们对于城市和生活环境的想法。这些想法将被纳入到设计概念中统一考虑。其次，城市设计概念由市民进行评估，公开讨论、改进，并最终定稿。接着，居民可以在自己的地块内建房子。这种方式得到了人们极大的支持，主要是因为这不是一个颠覆性的全新设计，而是将原有的保留元素融入到新的规划布局中统一考虑。这个设计通过新旧城镇建设的对比，增强了原先的场地记忆。

城市规划为当地人民创造了社会和经济相融合的条件。在一些由小地块组成的新区，小体量的房子紧紧相邻排布。在其他一些有较大房屋集中的区域，房子可能会更大、更独立。这种混合方式允许各种类型的人和企业建造他们自己的房子或设施。很显然，具备构建家园的能力，是社区发展的巨大推动力，并且能在现有的环境中增加美的价值。

作为住宅和商业设施的补充，设计者特别增加了文化和教育机构，来激发地区活力，带动客流。公共空间的设计利用了灾难中幸存的树木。重建工程耗时近10年，于2010年完工。

小地块中的小体量房屋
Smaller houses in a small plot

Roombeek街区
Roombeek neighborhood

Roombeek街区
Roombeek neighborhood

恩斯赫德 ENSCHEDE PRODUCT TOWNS

如今，Roombeek街区人气旺盛，每年都有大量的游客来访。这种自建房屋、全程无开发商介入的规划战略，吸引了荷兰其他地区的纷纷效仿。现在几乎所有新的开发都采用恩斯赫德规划模式。其中，最突出的城市是阿尔梅勒新镇。那里大部分住宅建筑都是居民自己建造的，没有开发商的参与。

灾难体现了小镇很强的恢复力。一个完美街区的建设，为整个国家的其他地区树立了榜样。相信恩斯赫德将继续前行展现更强的生命力。

编者感言

灾难，带给人们的往往比苦痛要多很多。其实，我们一直在思考的是，完美是不是就等于幸运。也许是幸运的，却也失去了看清浮沉苍生的机会。小到个人的命运，大到国家乃至人类。历史早就给了我们答案。人们总说，荷兰是一个童话般的国度。只是人们在诉说的时候总是忘却或刻意不提那些苦难和荷兰作为海上马车夫的血性和拼搏精神。

荷兰其实是一个任性十足的国家。荷兰人有着异想天开的设计思维，却也有果断勇敢的执行力。率真、追求美好，橙色的热情和乐观，也许才是这个国家留给世人的启示。有些我们觉得不可能的创新和尝试，他们都已经付诸实践。这让人不得不对荷兰设计师和政府有些肃然起敬。

童话里不只有美好，而是要拥有一颗经历苦难，经历悲惨世界后却仍然相信童话的坚韧之心。这样，就算世界毁灭梦想千百次，我们却还是可以拾起书写和描绘童话的那支笔。

图片来源：https://www.rijksmuseumtwenthe.nl

Roombeek街区的主要街道
Mainstreet Roombeek

科技小镇
SCIENCE TOWNS

来自荷兰的案例
Cases from the Netherlands

134/莱顿	LEIDEN
144/瓦赫宁根	WAGENINGEN
152/格罗宁根	GRONINGEN

莱顿地图
Map of Leiden

莱顿 LEIDEN　　　　SCIENCE TOWNS

莱顿维特辛格运河
Witte Singel Leiden

经典与新葩交相辉映的发现之城——莱顿

莱顿让人印象深刻的还是那所大学。如今想来，在有生之年，在灿烂的年华，那值得一生怀念的大学时光，也许还是有一点点遗憾。其实笔者是很贪心的，想要去很多大学，无奈没有赫敏的时间倒转器，不能一下子去许多优秀的大学学习。而莱顿的居民就很幸福了，走在城镇的小径上，到处都是大学的影子，花径中各个学院的图书馆不停地邀请着路人，许多设施也不需要学生卡。这才是便利和亲民的大学。至今还记得在大学图书馆的一些尴尬之事，不过，都被大度温柔的图书管理员给包容了。也许现在去读书，我们会选择这样的一座城市。大学即是城市，城市孕育着大学。

大学之道，在明明德，在亲民，在止于至善①。喜欢这样的大学，没有围墙，没有太多设限，就算没有学生卡，也是其中可以摘取知识的一个平民。

①出自《大学》原意为：大学的宗旨在于弘扬光明正大的品德，在于使人弃旧图新，在于使人达到最完善的境界。

莱顿学院楼
Academie-gebouw Leiden

1649年莱顿莱顿规划：贸易技术行政城镇
*Paln of Leiden Leiden 1649:
Town of trade science and administration*

1659年莱顿剪影：完整的荷兰城镇
Silhoutte of leiden 1659: Complete Dutchtown

莱顿 LEIDEN　　SCIENCE TOWNS

一座发现之城

莱顿是17世纪与代尔夫特齐名的荷兰共和国著名小镇，在这座城镇中，街道、广场、运河和400多年丰富悠久的学术历史有机地交织在一起。这座城市遍布着教育、文化机构等设施。如今，莱顿除了大学，还有如雨后春笋般涌现地生物医药科学中心。在荷兰兰斯塔德地区城市网络中，莱顿是当地居民，特别是有孩童家庭首选的舒适、安静、设施先进的栖息地。

莱顿和其他许多经典的荷兰小镇一样，地处两条河流旧莱茵河(Oude Rijn)和新莱茵河(Nieuwe Rijn)的交汇处。慢慢地因贸易、生产和管理而定居的人们，汇聚和居住在此处，形成了莱顿。从15世纪起，这个小小定居点逐渐合并、改造，形成了一座有城墙的城市。那里有荷兰最重要的水道。水道将莱顿和北海在卡特韦克附近相连，东部与乌得勒支相连，南部则与鹿特丹相连。由于地处贸易路线的交叉点，城镇蓬勃发展起来，莱顿成为了围绕亚麻生产的一个重要工业中心。此外，莱顿也在与1568～1648年的西班牙王国的80年战争中发挥了关键作用。在抵抗西班牙压迫的过程中，莱顿在战争早期的1575年10月3日解放。这对在威廉·奥兰治亲王(Prins Willem van Oranje)(也称沉默者威廉)统一荷兰而言是很小但非常重要的一步。后来威廉·奥兰治亲王被称为荷兰的"国父"，莱顿人民则每年都在10月3日庆祝纪念胜利。

莱顿没有其他典型欧洲城镇具有的广场和市场。其令人难忘的空间是旧莱茵河和新莱茵河的城市水道，欢乐曲线沿着历史中心的"莱顿城堡"(Castle of Leiden)流动。围绕河流交界处的北部是旧韦斯特(Oude Vest)，南部是石谷仓(Steenschuur)。石谷仓沿着莱顿最著名的街道拉彭堡(Rapenburg)，它蜿蜒地围绕着主教堂彼得教堂。彼得教堂是小镇中最古老、最宏大的建筑之一，其历史可追溯到1390年，采用晚期哥特式建筑风格。彼得教堂对面则是最著名的建筑——学院楼(Academiegebouw)。几个世纪以来，学院楼一直是大学的核心。最初奠基是在1516年，建筑被作为多米尼加人或白衣修女的修道院。该建筑由八个拱形组成，有颇高的玻璃窗，整个建筑都体现了荷兰的务实风格。建筑内最著名的空间是"辛勤间"，那里有刚毕业的硕士和博士的签名墙。其中亮点是前女王比阿特丽克斯和现任国王威廉·亚历山大的亲笔签名。历史上，有许多王室成员都曾在莱顿大学学习。

莱顿拉彭堡
Rapenburg Leiden

莱顿城堡
De burcht leiden

莱顿石谷仓
Leiden Steenschuur

1920年卡默林昂尼斯实验室
Kamerling onnes laboratorium 1920

莱顿 LEIDEN

SCIENCE TOWNS

1575年，在莱顿坚持抵抗西班牙军事侵略的奥兰治亲王设立了莱顿大学。这不仅仅是历史的馈赠，也是历史发展的必然，荷兰共和国将得益于受教育良好的公民。而那时候较好的教育机构都是在荷兰以外的地区。早期的大学成立于13~14世纪根特和布鲁日地区，即现在的比利时。莱顿大学的创立伴随着学术科学的兴起，是16世纪欧洲文艺复兴的产物。莱顿大学被尊为欧洲领先学术机构的时间并不长。早期的著名学者如国际法创始人胡果·格劳秀斯(Hugo Grotius)和最早的学术神学家、古典主义者之一的赫拉尔杜斯·伏斯(Gerardus Vossius)。另一位来自莱顿大学的杰出人物是赫尔曼·博尔哈夫(Hermann Boerhave)，他被普遍视为现代医学的创始人，并在莱顿拥有自己的博物馆。虽然大多数人不都知道这些学者是谁，但莱顿的许多建筑物、街道和机构都是以这些学者的名字命名，这使他们的名声得以广泛流传。

莱顿大学经历了兴衰，但其规模和声誉仍在稳步上升。19世纪末20世纪初，大学在物理和化学领域出现了许多著名人物，如荷兰物理学家范德·瓦耳斯(van der Waals)和洛伦兹(Hendrik Antoon Lorentz)，以及天文物理学家泽尼克(Frederik Zernike)。

另外这里也建立了世界上第一个大学低温实验室，海克·卡末林·昂内斯(Heike Kamerlingh Onnes)教授在这里取得了只比绝对零度的-273摄氏度高一度的温度。1908年他还成功在液化氢金属中第一次发现超导电性。像爱因斯坦、弗洛伊德这样的科学巨人在20世纪上半叶曾在莱顿大学生活和工作了很长时间。如今，莱顿大学被认为是欧洲大陆顶尖的大学之一。大学有大约25000名学生和3000名员工，其中就有16位诺贝尔奖得主。

在莱顿大学的发展过程中，其周围的建筑及机构也已经融入了莱顿市民的日常生活。城镇规划和大学的发展齐头并进。学生的出现给小镇带来了新鲜活力。许多令人印象深刻的学术机构已经与城市的持续扩张相融合。

早期的建筑都集中在历史中心区。在19世纪历史中心区外围的开发区，可以看到像Huizinga研究所、大学图书馆、大学民族学博物馆这样的大型经典建筑。幸运的是，莱顿历史中心附近的防御区仍然完好无损，完美展示了历史小镇中心的面貌。如今，以前的防御工程已被改造成一系列围绕历史悠久市中心的公园，有大量的拥有绿色步行、聚会空间的大学或机构建筑。莱顿选择保留防御区，而不是像其他城市那样建设内环路，因为设立内环路会增加车辆进入历史小镇和大学设施的难度。自20世纪初，莱顿的发展已大大超出了其历史边界，莱顿70%的设施在1945年左右建成。在第三波发展中，莱顿创造了新的大学区域，其中火车站北部的现代生物科学中心是最具代表性的。

莱顿生物科技园：奥尔特大厦
Leiden Bioscience Park: Oortgebouw

莱顿自然历史博物馆
Leiden Museum for natural history

莱顿生物科学园
Bio Science park Leiden

莱顿生物科学园：当代建筑
Leiden Bioscience Park: Contemporary Architecture

莱顿 LEIDEN　　SCIENCE TOWNS

莱顿生物科学中心拥有强大的商业研究和科学机构，它们彼此成就、相互促进。生物科学中心对莱顿大学生命科学创新和大型出版业经济起着关键作用。生物科技园是一个技术创新体，链接着世界主要经济区域。

莱顿科技园里有荷兰著名建筑师最令人印象深刻的现代建筑作品。UN设计事务所的本·范贝克尔(Ben van Berkel)设计了生物科学园区的跨国制药企业总部的未来建筑。埃里克·范埃格拉特(Erick van Egeraat)设计了物理和天文学研究所的新建筑。莱顿完美平衡了经典建筑和现代建筑的发展。继研究机构和学术建筑建造之后，莱顿又出现了博物馆建筑小高潮。这充分符合小镇的科学气质，为荷兰居民提供了充满教育氛围的休闲娱乐。莱顿最大的新博物馆是由欧洲著名的博物馆建筑设计师努特林斯·雷代克(Neutelings Riedijk)对自然博物馆进行设计重建的。这些博物馆扩大了以"发现之城"为口号的莱顿城市品牌。

重建大学的历史也值得回顾。人们不应该忽视这样一个事实：莱顿成为10多万居民的便利社区，其中很多人却与大学及其生态系统并不相关。可以说是莱顿的悠久历史、便利社区、丰富的设施和服务，以及作为一个大学小镇的特色，吸引着人们在那里生活。而实现这些的是靠近A4高速公路和服务于兰斯塔德大都市经济动脉的铁路交通系统。住在莱顿，30分钟内可以到达鹿特丹、海牙和阿姆斯特丹，15分钟内乘火车或汽车可以到达史基浦国家机场。"市长和教授区"(Burgemeesters and professoren wijk)是这个社区集中体现的特点。社区始建于1930年，靠近城镇历史中心。宽敞的布局和优质的建筑，接近中心设施，交通便利，使其成为了深受欢迎的家庭生活区。莱顿特点是其小规模的历史生活环境中的高便利性，距离荷兰所有的主要经济中心的距离都较短。可以说，莱顿是地区或国家范围内吸引人的城镇，拥有舒适和熟悉的环境。相对于较大的城市，尤其是有孩子、需要安全舒适生活场所的家庭，是一个不错的选择，其生活成本也明显低于阿姆斯特丹那样的大城市。

莱顿的便利生活：教授居委会
Convenient living in leiden : Professorenbuurt

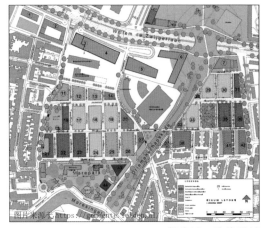

新莱顿：总体规划
Nieuw Leyden Leiden: Masterplan

新莱顿：城镇中心的住宅重建
Nieuw Leyden Leiden:
Residential redevelopment in town center

新莱顿：城镇中心的住宅重建
Nieuw Leyden Leiden:
Residential redevelopment in town center

莱顿 LEIDEN　　　　　　　　　SCIENCE TOWNS

吸引前卫居民最有趣的当代建设项目是叫新莱顿的社区，一个位于城市历史边界内的高密度低层街区。城市设计是基于一个网格。莱顿市民可以享受很多特殊利益，也可以建造他们自己的房子。在这个实验性的街区，居民可以创建自己的家园，不过个人房屋的建设受到总体规划的约束。最后的指导方针和总平面设计图则是由荷兰著名设计公司MVRDV提出的。

这个方案提供了一个和谐的发展思路，其中每栋单独的建筑物都能为整体的美丽和健康发展做出贡献。由于大部分房屋是小天井排屋，从而形成一个较高密度的居住区。建筑风格都遵循进步建筑的形式，采用天然材料，细节精致。此外，也设计通过废物和能源流动的循环方法实现可持续发展。莱顿这种发展模式代表了荷兰发展的趋势。公民可以根据城市大环境建造自己的房子。越来越多的人热衷于这种方式，因为他们不需要从开发商手中购买已经规划好楼层和立面的住房。而因为近期的经济危机，大规模的资本密集型项目在银行危机期间被搁置，更加速了这种以市民为基础的有机城镇发展模式。

新莱顿自建房屋的模式定义了家庭生活的新类型，并将鼓励其他城市的人们跟随他们的脚步。

历史悠久的市中心，美丽的运河，新潮的博物馆，自我建设的社区，大学和其生态系统，一起创造了一个优雅、充满个性的中心服务城镇。莱顿，一座发现之城，让人们发现他们不一定需要住大城市。

编者感言

大学和城市，怎样才能更自然地融为一体，莱顿已经给了我们很好的答卷。也正是有那么优秀的大学，莱顿才开启了科技小镇的规划建设。莱顿很好地利用了大学本身的学科和学院优势，也吸引更多外来人口和企业的驻扎。

其实，很多年前，国人就在不停讨论一个问题，国外的大学是没有围墙的。这其实不仅仅是物质上客观的现象，更是有些灵魂深处的精神体现。

University也好，大学也好，这两个词语其实涵盖的远远比一所学校要多的多。然而，在世界上的许多例子中，一座大城市往往有很多的著名大学。这往往也造成了激烈的竞争，更有甚者，也会为了大学之名争破头皮，大开骂战。

不得不说，莱顿是幸运和幸福的，没有这些背驰大学精神的杂音。学校，许多时候是一种不可磨灭的印记，它将伴随我们一生。许多人、单位、机构，也许只是看到你拥有这个印记的光环。但其实，大学锻造的是你的灵魂。多年以后，这种体会和共鸣会更加强烈。

瓦赫宁根地图
Map of wageningen

瓦赫宁根 WAGENINGEN　　SCIENCE TOWNS

瓦赫宁根自然风光
Wageningen natural scenery

欧洲环境、农业、食品的大脑——瓦赫宁根

用现在年轻人的话来说，瓦赫宁根就是非常的高大上。科技的前沿，各种领域名列前茅。穿着白大褂，做着科研的诸多科学家、学术精英。虽然很多时候他们不为人们熟识，但却是实实在在地为人类做着贡献。

世界的未来之一就包括生物科学、环境科学，这是人类发展到一定阶段必须审视和重视的问题。瓦赫宁根也许就是未来城市发展的雏形。科技，应该是人类未来的核心。如何让科技更美好地服务于人类，并让人类融于自然，也许正是未来城镇要解决的主题核心。不如展望一下，科技可以拯救更多的人，也能让人类超脱出一些局限。这些进步许多是里程碑式的，毕竟我们面对的是整个地球乃至更大的宇宙。

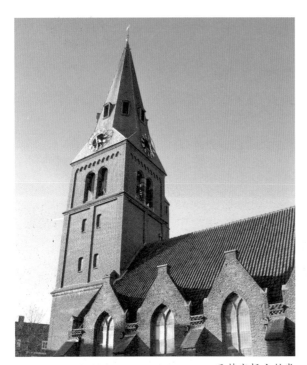

瓦赫宁根主教堂
Main church of Wageningen

古城的城堡和城墙遗址
Old city ruins of the castle and city walls

点燃自由之火是瓦赫宁根自二战胜利以来的传统
Lighting the fire of freedom is the tradition of Wahningen since the victory of World War II

世界酒店
Hotel de Wereld

瓦赫宁根 WAGENINGEN

瓦赫宁根是一个位于莱茵河畔的典型荷兰滨水城镇。地理位置成就了瓦赫宁根独特的沙地和河流景观。自19世纪以来，农业学校的存在已成为其最显著的特征。如今，瓦赫宁根大学及其许多世界知名的研究机构已经成为健康、农业、食品和生活环境的专业研究中心。当然小镇还为居民提供了舒适、安静的居住环境。

瓦赫宁根位于荷兰中部海尔德兰省(Gelderland)，地处莱茵河畔，面积32平方公里，人口3.7万，是欧洲甚至世界著名的"生命科学之城"。

瓦赫宁根是荷兰的历史名城，有中世纪早期的教堂和林场遗迹。瓦赫宁根堡和斯特拉堡是城镇的主要城堡。1263年，瓦赫宁根取得城市立政权。1918年，瓦赫宁根市创办了第一个教育机构——兰布华格学校，该校即瓦赫宁根大学的前身，由此开始了这座城市由古老小城发展成现代农业科技园区和食品谷中心的历程。

瓦赫宁根因历史事件而闻名。1945年5月5日代表盟军的查理斯·福克斯(Charles Foulkes)将军和德军司令约汉尼斯·布莱科威茨(Johannes Blaskowitz)在瓦赫宁根的"世界酒店"(Hotel de wereld)就德军投降进行了谈判，并签订了德军投降书。对荷兰来说这意味着第二次世界大战的结束，所以瓦赫宁根又被称为解放之城。这也是其他欧洲国家的选择了5月8日为二战胜利日，而荷兰却将定为二战胜利日定在5月5日的原因。

众所周知，荷兰的农业居于世界先进水平，荷兰土地总面积的56%用于农业。全世界任何一个国家的农业在高效、多产以及创新能力上都比不上荷兰。打个有趣的比方，假如全世界所有国家的农民都像荷兰农民这样高效多产，那么只需要目前全世界农业土地的1/4就能解决粮食问题了。荷兰的奶牛产奶量高，但由此产生的温室气体排放量却低于其他国家；化肥使用量越来越少，因为能从不同的动物粪便中提取宝贵的原料；荷兰不再需要依靠进口磷酸盐等，都是荷兰先进农业的写照。

荷兰农业的先进可以追溯到黄金世纪。由于当时大多数粮食必须来自波兰，荷兰农民被迫专注于农业生产，他们极力将科技运用于耕地和农业中，最好的例子就是耕地的排水以及土壤肥力的提高。这些技术后来从荷兰传播到欧洲乃至全世界。

根据世界贸易组织和荷兰中央统计局统计的农产品出口总额，2015年荷兰农产品出口总额达813亿欧元，创下了历史记录，居全世界第二，仅次于美国。如果加上农业相关产品，如农业机械以及肥料，出口总额达到900亿欧元，占荷兰全年出口总额的21%。一个占地面积只有4万平方公里的小国家，农产品出口总额竟然是全世界第二，显然不是依靠本土的农产品出口。这主要归功于荷兰农业的进口后加工再出口的成熟贸易，而这种贸易依靠的是强有力的高科技支持。

瓦赫宁根大学校园全景
Panorama of the campus of Wageningen University

瓦赫宁根大学礼堂
Aula of Wageningen University

瓦赫宁根大学阿特拉斯研究大厦
Atlas building WUR.

瓦赫宁根 WAGENINGEN

而在荷兰，提起农业高科技，就会直接想到瓦赫宁根，想到瓦赫宁根大学。除了历史事件以外，瓦赫宁根还以拥有致力于生命科学教学、科研的瓦赫宁根大学而闻名，同时瓦赫宁根也是食品谷的中心城市。可以毫不夸张说，瓦赫宁根之所以成为荷兰国际化的城市之一，很大程度上来源于瓦赫宁根大学的国际化；瓦赫宁根之所以成为生命科学之城，很大程度上来源于瓦赫宁根大学的生命科学研究成果。

瓦赫宁根大学（全称：瓦赫宁根大学及研究中心，简称：WUR），是荷兰14所U类大学（研究型大学）之一，在荷兰高等教育指南上连续九年高居榜首，在生命科学领域亦是欧洲和世界的领头羊之一。它是唯一一所由荷兰经济部直接拨款的大学，不同于其他由荷兰教育部拨款的大学。瓦赫宁根大学的专业设置涵盖了从动植物科学到食品环境科学等诸多领域，还包含有景观设计、经济、发展研究等非自然科学专业。其中最负盛名的当属植物科学、环境科学、食品科学等。景观设计、国际发展等专业也实属优秀。综合而言，瓦赫宁根大学属于一所非常有特色的研究型大学。

瓦赫宁根大学及研究中心由瓦赫宁根大学、万豪劳伦斯坦农学院以及农业科研机构组成。它始建于1876年，是荷兰最早提供农业教育的学院。在随后的近40年里，不断发展壮大，并于1918年正式被荷兰政府批准成为学术性大学，更名为荷兰国家农业大学。1997年与荷兰DLO农业研究基金会旗下科研机构合并，合称为瓦赫宁根大学及研究中心。"探索大自然的潜力，提高人类生活品质"（To explore the potential of nature to improve the quality of life）是瓦赫宁根大学及研究中心的教育和研究宗旨。近百年来，瓦赫宁根大学及研究中心一直致力于生命科学及自然资源领域相关的高等教育及科技研究，在国际上享有很高的名誉和声望。

在食品安全和食品质量方面，瓦赫宁根大学及研究中心启动和参与过的欧盟项目高达31项（共80项），充分显示了其在这一领域的实力和国际地位。在农业方面，瓦赫宁根大学及研究中心的文献引用量在世界范围内排名第三。在植物和动物科学方面，该大学的文献引用量世界排名第五。农业技术方面，瓦赫宁根大学为世界不同地区设计不同的温室系统，将高效经济生产与最少的能源、水和营养物投入相结合，提高全球各地农业的生产力，同时瓦赫宁根大学独特的生物防控技术，即利用天敌来对付植物病虫害的技术，为保障食品安全和环境保护的研究领域开拓了新的方向。环境科学方面，世界人口与财富的增加使得资源尤其是水资源的竞争剧烈。瓦赫宁根大学对节水农业的开发和应用领域也做出了卓越的贡献。

瓦赫宁根农业科技商业园区(Agro Business and Science Park)集中了大中小型有关生命科学，食品以及健康领域的各类公司和研究机构，科研人员达8000多人，加上瓦赫宁根大学的科研支持，被誉为"欧洲的生命科学中心"。

食品谷（Food Valley）是荷兰食品及营养研究集群的所在地，由许多食品、营养领域的研究机构和代表目前先进工艺水平的企业组成，被认为是当今世界上最大的食品营养研发集群。

Lumen大厦
Lumen building

Lumen大厦的温室的全景
Panorama of the greenhouse of the Lumen building

夏日的校园
Summer on Campus

瓦赫宁根大学农业科研
Wageningen University agricultural research

瓦赫宁根 WAGENINGEN

SCIENCE TOWNS

从1972年开始，瓦赫宁根大学研究中心每年会为荷兰政府提供一份"农业经济报告"，这是一份关于荷兰农业的经济状况调查报告，内容包括世界整体农业经济和政治发展方向，欧盟农业政策的变化调整，荷兰农业综合体和食品工业的发展，荷兰农村地区的变化和环境问题，农业生产结构和生产要素的描述以及农业相关子行业的盈利能力等。此报告会为荷兰政府提供农业战略的综合指导。

显而易见，瓦赫宁根大学的科研优势是瓦赫宁根的核心特色，带给了瓦赫宁根独一无二的国际形象，在此基础上，凭借农业科技商业园区和食品谷的平台，构建了围绕生命科学的主导产业，成功吸引了来自全世界的相关企业落户瓦赫宁根。而这些企业来到瓦格林根，除了瓦赫宁根的国际形象以外，瓦赫宁根大学源源不断毕业的大学生也为企业提供了优质的人力资源，瓦赫宁根大学的科研优势为企业提供了最新的科技资源。这是一种相互依赖又相互吸引的互惠互利的双赢关系，超越了传统意义的招商引资，是知识和市场经验的完美结合。另一方面，企业的落户又给瓦赫宁根的毕业生提供了更多更好的就业机会，不仅私人机构，许多政府机构也选择驻扎于此，比如荷兰国家动植物检验检疫局就设在瓦赫宁根内。

为了增进企业家之间的信息交流与技术合作，相关机构会组织在瓦赫宁根最著名的世界酒店举行每月一次的交流会。这种交流会尤其为那些怀揣梦想落户瓦格林根的年轻企业家提供了良好的创业创新生态环境。

瓦赫宁根的文化生活极其丰富，城市国际化程度很高，随处可见海外学生。大学建筑集中在城镇四周，而学生公寓则在城内，瓦赫宁根的Y型（又称星型公寓）学生公寓因其不寻常的形状以及坐落在城镇边缘的地理位置给瓦赫宁根的天际线增加了一道独特的风景。

瓦赫宁根既是历史名城，也是生命科学之城，更是一座充满生机之城。

编者感言

科技的力量往往是不可估量的，瓦赫宁根对世界的贡献不言而喻，它已经代表了尖端科研领域的应用性尝试。走在这样一座城镇，总觉得该是肃然起敬和分外虔诚的。只是它所透露的和谐和宁静却又顿时让人减轻了心理压力。更好地利用自然，合理种植，科学实验才是这座城市秉承的真理。所以，它总是到处透露出人与自然的和谐理念，也充盈着人类对万物的感恩之心。

是的，在这样一座充满正能量的城镇中，你可以自由地呼吸、欢笑，而更多地却是被晕染上了一层认真和细致。这种责任感，是和科学一样严谨的。作为茫茫人海中的一员，你是不是也该去尽力发挥自己的应尽之力。

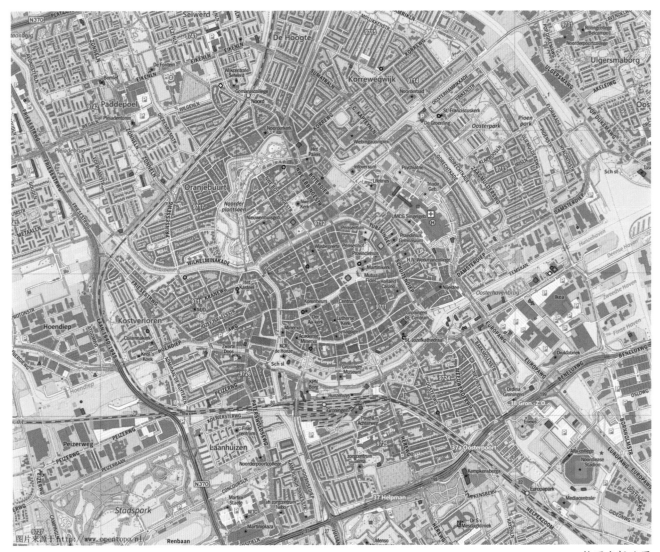

格罗宁根地图
Map of Groningen

格罗宁根 GRONINGEN　　　　SCIENCE TOWNS

格罗宁根中央广场
Central Square of Groningen

亦庄亦妍，恒久弥新——格罗宁根

格罗宁根也是一座和大学融合良好的城镇，但又不仅仅只依托于大学而进行城市开发。这是一座更加多元化的城镇，拥有现今流行都市都具备的元素。所以浓缩城市这个题眼颇为符合，不过它又不同于摩登大都市。它也保留了属于荷兰特有的那份历史痕迹及气息。可以说它是一个很好的融合了现代化和文艺气息的城镇。

这应该是很不错的城镇发展范本，拥有摩登，却也富有诗意。所谓的左手执羽，右手执干。能文能武，舞蹈俱佳也。

格罗宁根老城
Groningen Old Town

1649年格罗宁根历史地图
Historic map of Groningen 1649

1659格罗宁根剪影：防御城墙后的城邦
Groningen Sihouet 1659: City state behind defense Walls

格罗宁根 GRONINGEN

SCIENCE TOWNS

格罗宁根是荷兰兰斯塔德外围地区最有活力的城市,拥有20万左右人口。几个世纪以来,格罗宁根一直是北荷兰无可争议的首都。作为一个历史中心,格罗宁根综合了贸易、行政生产和教育的功能。格罗宁根大学是世界百所名校之一,拥有国际学生3万多人,其中许多学生和教员都是中国人。在城市设计和城市布局方面,格罗宁根具有20世纪具有代表性的城市布局类型,有购物中心、大学区域、与周围的开放景观相融合的可步行和车行的环路。

现在,如果在工作日探访格罗宁根,人们可以在充满历史感的街道上感受学术氛围。午餐时间,城里到处是年轻人和学者。他们找地方吃饭、会谈,准备参加下一场讲座或实习会议。靓丽学生们的活动都在鱼市(Vismarkt)和大广场(Grote Markt)周围的历史场景中展开,指引着几个世纪以来一直存续的诸如马提尼教堂(Martini Toren),科伦比尔斯(Koren-beurs)建筑和学院大楼遗迹。一天结束时,小镇又恢复了活力,学生和教员们纷纷奔赴街道和街区,参加他们的晚间约会。几个世纪以来,格罗宁根做为区域首府,在其有限的空间内却容纳了所有大型城镇拥有的设施。格罗宁根小镇就像一个充满自信的学者,虽然经历了很多,但总是知道如何适应新的环境,与时俱进。

有人称格罗宁根是一个微型的都市,因为它具备了现代生活服务的综合特征,从博物馆、办公楼、经典遗址,到环境融洽的商场和大型足球场。漫步在格罗宁根,人们会觉得街道无限地从一侧连接到另外一侧,而前方总会看到美丽的运河或公园。在这里,新老交汇、古今融合。精致的玻璃建筑矗立在古典建筑旁。这种友好、和谐的尺度,营造了一种温暖的舒适感,从中央广场延伸至周边空旷地的牧场,再到城郊区域。

在学年伊始,所有的新学生都会庆祝当地的"Bommen Berend"节日。这时,格罗宁根的历史显得格外突出。这是庆祝1695年小镇解放的节日。那时候,德国称为Berend的明斯特主教率先使用大炮和炸弹,试图突破荷兰共和国第七省北部首都的抵抗。格罗宁根在严酷的考验下没有被削弱,而是成长为一个有韧性的小镇,成为贸易、行政和教育的热点地区,统领着北荷兰省。

秋天的格罗宁根自然博物馆
Autumn Groningen Nature Museum

斯塔茨运河
Peat canal Stadskanaal groningen

1894年格罗宁根沿着运河建造的Ubbo emmius街道
Ubbo emmius singel along the canal. Groningen 1894

格罗宁根 GRONINGEN SCIENCE TOWNS

格罗宁根大学成立于1614年，最早是神职人员的学校。这是许多大学普遍的来源，因为宗教在当时社会中起着核心作用，神学讨论对政府和行政也会产生重大影响。早在17世纪，已经有很多海外留学生，为大学增强了讨论和创新的开放性学术氛围。这种文化使格罗宁根大学发展成为如今的全球百强名校之一。名为学术楼的主要行政大楼，至今仍矗立在小镇的中心。其他设施，如礼仪学院楼、大学董事会和宾馆，则都位于老城中心。令人印象深刻的学术医院综合楼，位于老城边缘，在城市防御墙拆除后建立。老城北边是泽尔尼克(Zernike)公园，科学和经济学院设立在那里。所有建筑之间骑自行车就可以互达。格罗宁根大学完全融入了城市，与邻里街道交织在一起。教职人员、学生和城市的融合，是格罗宁根活泼、友好的城市氛围的关键。大学和城镇的融合是格罗宁根最具特征的地方。这是格罗宁根的特色之一。而不像如今的许多大学校园都与城市及中心隔离。

走在格罗宁根，看着那些倾注心血的建筑，人们忍不住会想格罗宁根的人们需要倾注多少时间和金钱来创造这些不朽的建筑。例如老城区鱼市场旁边的科伦比尔斯，就是前100座新古典风格的代表之一。在建设时期，因其古典对称的立面和精致的内部钢结构，被认为是前沿建筑。它是粮食贸易的中心，也是城市及周边地区的重要财富来源。格罗宁根北部盐渍黏土是几个世纪以来用于供应荷兰大部分地区的粮食及其他农产品生产的沃土。

格罗宁根财富的另一个来源是小镇东南地区泥炭开采的垄断。这个被称为"veenkolonien"（泥炭群落）的开发地区是通过运河与格罗宁根市相连的。在城市中，泥炭被用作燃料，并出口到荷兰其他地区。该地区泥炭经济的地理印记仍然随处可见。富商们沿格罗宁根运河建造的房子，都是通过泥炭开采得来的费用建造的。格罗宁根有许多水道，例如斯塔茨运河(Stadskanaal)，都是开挖用于泥炭运输的。格罗宁根附近的泥炭群落景观（目前主要的旅游目的地）也是受到泥炭的重大影响。草皮开挖后，土地被用于农业和作物；随后被租给农民，他们对城市财政做出了重大贡献。很多过去的工业基础设施，现在变成了格罗宁根的特征区域。

格罗宁根风光
Groningen scenery

格罗宁根避风港：从前的市中心过渡港口
*Noorderhaven Groningen:
Former transition harbour in the Town center*

格罗宁根大学主楼
Academie gebouw groningen

格罗宁根Ubbo Emmiussingel街道：19世纪的延伸
Ubbo Emmius Singel Groningen: 19th century extension

格罗宁根 GRONINGEN   SCIENCE TOWNS

格罗宁根小镇中心附近有许多小尺度、步行可达的社区和低层建筑物。有些地方偶见教育或更大规模的机构。有个神奇的社区Hortus kwartier位于大广场北部。Hortus地区以植物园Hortus botanicus命名。它的街道和建筑按十七世纪的小型街道和深街区进行布置。在人们以船只及后来的火车作为交通工具，但未使用汽车前，单体住宅和建筑物有机发展的小规模组团模式设想就已形成了。这种小尺度、可步行的迷人环境，在今天自然更加受欢迎。这个社区里有一种特殊的住宅类型：古老的围屋，很多住宅围绕着一个规划好的花园，并有一扇门通向街道。在古代，这些院落式房屋住的都是寡妇或单身妇女，有私人的户外空间。如今，这些房屋深受喜欢安静、靠近老城区的学生和年轻夫妇的欢迎。

走在格罗宁根，你会在不同的历史区域感受到历史的层次感。小镇中一些很棒的街道，例如Ubbo Emmius Singel, Herensingel和Noorderplantsoen街道，与闹市区的感觉截然不同。这种19世纪后期的发展主要得益于关注资本主义的资产阶级，由此造就了很多美丽的房屋。这些开发主要集中在市中心边缘地区，原先城市防御墙的位置。19世纪末期，荷兰政府决定重新规划国防有关国家基础设施的问题。因此，荷兰所有城镇的防御墙都是根据1874年《国防法》被拆除的。环状城市防御工事的拆除，以适应于20世纪城市的现代化和扩展。

古老的庭院式住宅
The old fashioned yard house

Hortus botanicus植物园
Hortus botanicus

格罗宁根图书馆：精心整合的现代建筑
Library Groningen:
Carefully integrated modern architecture

格罗宁根Waag：新老建筑
Waag Groningen:
Old and new architecture

Brinkflats：雷姆•库哈斯的第一栋建筑
Brinkflats Groningen:
First Buildings of Rem Koolhaas

格罗宁根大学医学中心：当代建筑
University medical Center Groningen:
Contemporary Architecture

格罗宁根 GRONINGEN SCIENCE TOWNS

格罗宁根小镇以其实验性的、精湛的城市规划策略而闻名。在20世纪，适度增加了许多新的城市规划策略，与城市中较古典的部分相结合。这种方法有助于保持步行街道，以及大小建筑之间的连续性，从而不破坏20世纪消失的"有机城镇的感觉"。即使在历史小镇的中心，当代建筑也能与过去的经典联系在一起。格罗宁根在住宅聚集区和改造区应用新的方法，具有悠久的历史和经验。一些著名建筑师和前卫设计师的参与，促进了该地区的蓬勃发展。

城镇建筑发展的一个例子便是是位于Verbindings kanaal运河交界处的边缘公寓(Brinkflats)，是著名的荷兰建筑师雷姆•库哈斯(Rem Koolhaas)在1988年实现的他的第一个总体规划和公寓项目。另一个著名的例子是大市场附近的Waagstraat综合体。那是意大利建筑师阿道夫•纳塔利尼(Adolfo Natalini)在大众评审的竞赛中脱颖而出的得意之作。由大众评审选出建筑和城市设计，是格罗宁根早期的标志政策。往往由专业规划师作选择时，就会有不同的结果。此外，意大利建筑师乔治•格拉斯(Giorgio Grassi)设计的位于Oude Boteringe街的人民图书馆，是一个在现代建筑中融合经典形状的、奇妙的人文建筑。这个相对较大的建筑入口分为三块区域，每块都类似格罗宁根经典房屋的大小和比例。中间留有一些区域用于图书馆门口的小广场。材料方面，砖与周围环境相匹配，但颜色和细节应用更倾向于20世纪的方式。设计与现代材料和细节融合的比例相似，使其与周围的经典风格达到平衡，并有对比。

城镇中现代建筑最重要的作品之一就是格罗宁根博物馆。它坐落在Verbindings kanaal运河的岛上，博物馆是从火车站到市中心步行路线上的重要节点，由三个展馆组成，外观小巧而通透。整个项目由意大利建筑师亚历山德罗•门迪尼(Allesandero Mendini)负责，每个展馆由不同的建筑师设计。博物馆展馆融合了理性主义的意大利建筑风格，亚历山德罗•门迪尼的孟菲斯设计风格和奥地利公司蓝天组(Coop Himmelblau)的解构主义建筑风格，建成后享誉世界。博物馆本身是后现代建筑的杰出例子，通过一种极端、不寻常的方式带入。因此，博物馆成了19世纪80年代国际建筑的图标。人们可以看到博物馆雄心勃勃的规划和精致的布置，这也见证了历经几十年衰落之后复兴的城市中心。在闹市衰退的过程中，居民和公司都离开了市中心。这对大多数战后郊区化的西方城市产生了深远的影响。

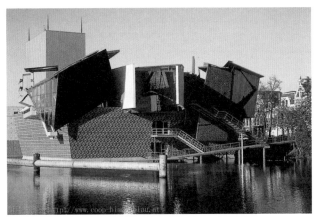

格罗宁根博物馆
Museum Groningen

格罗宁根博物馆夜景
Night view of museum Groningen

格罗宁根论坛：城市的新地标
Groninger Forum: New Landmark for the City

格罗宁根 GRONINGEN　　SCIENCE TOWNS

现在格罗宁根因其学术优势和顶尖人才，吸引着国际化人群。小镇因其历史和农业背景，并依赖其在北荷兰的区域中心地位，越来越重视服务业、国际合作和娱乐业。市中心变成了高规格活动和机构的热点区域。附近的绿树成荫的郊区非常适合成长型家庭。沿环路，商业和服务设施齐聚。郊区农村则是放松心情、享受生活的完美去处。

在2019年，由荷兰建筑师设计的格罗宁根论坛大楼将交付。这是一个服务于市中心的文化集群的独特前卫建筑。该建筑群包括了图书馆、展览中心、电影院和娱乐场所，形成垂直的组织结构，而在屋顶上有个视野开阔的观景台。此处会发展为城市中心的制高点，并将格罗宁根定义为21世纪面向全球、区域定位的科学休闲服务小镇。

编者感言

格罗宁根倒是真正的建筑设计前沿阵地，它的城市建筑及规划选择方式尤其让人印象深刻，不是政府拍板也不是专业人员界定，而是由大众评审得出。

这个其实不只是大胆而已，它也代表了当今世界的一种派别趋势。许多领域其实都面临着这样的冲撞和选择。许多时候大众喜欢的并不一定是专业人员选择的最佳。所谓的最佳畅销奖和专业类奖项的区别就在于此。不过，这也是我们需要深思的一个领域，受欢迎虽然有时候源于好的营销炒作，利用了驱众心理，但相信像建筑这种直白呈现的作品还是值得进行这方面的探讨和尝试的。美的东西带给大家的共鸣应该是有目共睹的。

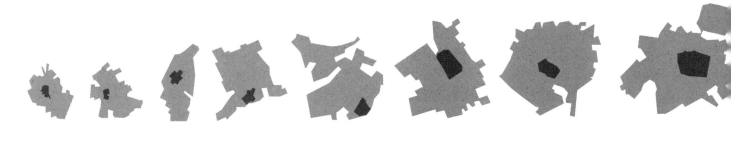

Laren Bergen Lisse Wageningen Volendam Delft Alkmaar Leiden

FEATURE TOWNS
Cases from the Netherlands

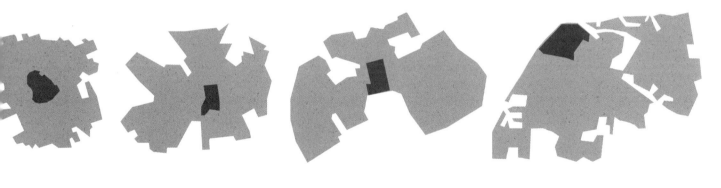

Groningen   Enschede   Almere   Dordrecht

FEATURE TOWNS IN HOLLAND

What makes a feature town?
"What makes a feature town and what makes a feature town in Holland?"
These are the central questions in this book. By analyzing different kinds of famous Dutch towns we aim to provide inspiration for feature town design in China that will be one of the main challenges put down by the government for the coming years. At the same time this book is intended to provide guidance and understanding for those urban professionals who plan to visit Holland.

Small country: global presence
The Netherlands is a small country in the European delta area along the North Sea. Yet the global footprint of the Netherlands in economic and cultural terms is surprisingly much larger than one would expect, given its small size and modest number of inhabitants. The wealth and global presence of the Netherlands has long historic roots. All historians agree that the identity and culture of the Netherlands are strongly related to the early Dutch condition of a network of small highly developed towns interconnected by navigable waterways. It is still this network of small towns shaped in the 16th century that forms the base for the power of the Dutch economy and the Dutch quality as it is today.

Modest scale
In comparison with Chinese cities and Chinese conurbations the Dutch network of main cities has a very miniature dimension comparable to the size of Shanghai. Capital Amsterdam counts 779.000 inhabitants and furthermore only Rotterdam and Den Haag also count more than 500.000 inhabitants. In a country of 18 million people there are only 30 cities and towns with more than 100.000 inhabitants. Yet each town has a carefully crafted balance of living and working, a diversified urban economy, facilities for leisure, cultural tourism and strong ties with the surrounding landscapes.

Dutch towns
Almost all Dutch towns have centers that are hundreds of years old dating from an era without automobiles, computers and airplanes. History is never far away and the Dutch are still close to the physical living context of their ancestors. Historic downtowns are still extremely popular for living, working and leisure, given the intimate human scale of these areas. Of course Dutch towns have been expanded and updated with their historic cores as points of departure. Modernization and expansion took command regarding water infrastructure, car infrastructure, industrialization, the public spaces and the individual buildings. Conversely, streets and canals of eras long gone have regained significance as decors for contemporary life and as habitats for players of a globalized economy.

Product of social and economical development
Dutch towns are known for their clear order and well defined structural concepts. Yet it is hard to describe the shape and size of the Dutch towns purely as products of planning and design. One needs to have a basic grasp of social and economic history and major events to understand towns' evolutions. The beginning of most towns relates to the conditions of local soil, waterways and the relation with the hinterland. Different historic periods extended and defined have made the shapes and urban constellations of Dutch towns as we can visit them today; towns with historic structures and additions of living, working, infrastructure and recreation. They can be understood as organically layered fabrics defined by incremental development of mostly private developments engaged within public frameworks.

History alive
Some towns have significantly expanded into complete cities active in the global economy. Others have kept a modest size and may have lost the significance of their heydays. But both have all the traces of history that are still there today in the form of buildings, streets, harbors, rivers and water systems. Most Dutch towns draw significant amounts of tourism based on their historic outlook, their historic values and historic preservations. Leisure tourism based on attractive historic features is an ingredient of all local economies.

A landscape of complete cities and feature towns
One of the urban trends in the 21st century is a split between larger cities that form nodes in the globalization of economy and culture versus the towns with or without special features that have a smaller scale and a more local focus. In Netherlands this process is well visible as it is in China, where the mega cities have international orientation and faster speed than secondary or tertiary cities. In this book we will not focus on the larger and complete cities but specifically focus on the Dutch feature towns; the relatively small towns with strong features that keep them vital. This will enable comparison with their future Chinese counterparts.

Through the lens of history
In order to get inspired and to understand contemporary Dutch towns we will look at them through the lens of history and analyze their contemporary outlook based on economic activity, urban planning and

architectural highlights. This book starts with a brief overview of the classic Dutch towns in the golden century and their subsequent expansion and industrialization into the 20th century. Then we will outline the challenges for the Dutch towns in the 21st century. The main body of the book is a series of 15 case studies of towns categorized in 5 distinct typologies on different scales.

THE CLASSIC DUTCH TOWN AND THE GOLDEN CENTURY

The abundance of all that can provide joy, partly because of the import of goods brought along the waterways and partly because of the fertile soil of the land. With the rivers full of fish and the efficient waterways surrounded by lush meadows, they say no other country has so many cities within such a small territory. The adages of erasmus, cambridge 1964, page 211.

Network of special towns in a flat land

After the Middle Ages a network of strong towns developed in the estuary of European rivers Rhine and Maas in an area that would be called the Netherlands. The swampy flat land in the delta accommodated an intricate web of smaller and bigger waterways. The interconnectivity of Dutch towns by easily navigable waterways was key to the early development of trade, and therefore the development of trade towns.

Feeding the town

One of the preconditions for the rise of the Dutch towns was a reliable system of agriculture that enabled the commercial production and trade of food and products on a scale that transcended rural and autarkical forms of settlement. For the rise of the Dutch towns a strong relationship between the town and the surrounding country side was necessary. Different kinds of food and products would be distributed to nearby towns and other regions on boats from the surrounding villages.

Early urbanisation

The Dutch towns had common characteristics in aspects of defense, administration water-management, trade, agriculture and religious beliefs. Yet it was usual for towns to develop identities related to the local context, the local infrastructure and the local skills of the people, resulting in unique products and unique services. Dutch towns developed specific identities that live on until today. In this era the Netherlands became the most urbanized area in the world with almost 50% of its people living in cities.

The golden century

Facing the same Spanish enemy during the 80-year war in the 16th century the Dutch towns and provinces assembled themselves to form the republic of the Netherlands. In the West part of Holland the republic was carried by a configuration of strong powerful towns that we still know today; Amsterdam Leiden Haarlem Rotterdam Delft Dordrecht and Enkhuizen. The republic of the Netherlands transformed into an unequalled platform for trade, transportation and warfare over water. This period made the Netherlands wealthy, cultured and the most powerful country in the world .Nowadays the seventeenth century is called "the Golden Century".

Global trade

The Dutch network of sea-faring water-towns took their production skills, trade expertise and sea faring abilities to a global level. One of the main accelerators in this transition was the VOC (Verenigde Oostindische Handels Compagnie). A company founded by the government for the lucrative trade between Holland and Asia. It was not only the first multinational company in the world, but also the first company in the world which shares could be publicly traded. Also the early financial innovations contributed to the rise of the Dutch empire.

Man-made country

Since a significant area of the Netherlands is built below the sea level, the Dutch like to say that their land is "borrowed from the water". To survive and prosper in this condition a refined water management technology with artifacts like dikes canals windmills evolved. To operate this system a high degree of organization and cooperation was necessary in between the towns, the government, the farmers and the landowners. Many historians have argued that the collective organizational skill derived from water-management and keeping the feet dry was another essential precondition for the rapid rise of the Dutch empire.

Merchant towns

During the golden century the trade-oriented water towns came to full blossom, with Amsterdam and its canals as most famous example. The integration of water transportation and urban design was, and still is, exceptional. The boats for transportation of goods could be docked in front of the merchant houses. These houses served as storage, showroom, office living quarters and family homes simultaneously. This setup of merchant houses along the commercial waterways facilitated global trade connections between individual merchant houses in the city and all the other port-cities in the world. This modus of urban design and waterway connectivity created an unparalleled network for trade opportunities that can be compared with the potential of the internet in the 21st century.

Landscape exploration

Profits made with overseas trade were reinvested in the Dutch towns. Large investments were made in town extensions around the canals but another effect was the colonization of the surrounding landscapes. In the seventeenth century the rich Amsterdam Merchants invested in the first large scale polder called the Beemster North of Amsterdam. The land in the polder was leased out to the farmers who provided Holland with food. This polder of almost 70 Km, designed with renaissance ideals of aesthetics was a stunning feature of early engineering skills. This new manmade landscape had a beauty that mankind had not seen before. Today the Beemster is a world heritage site assigned by UNESCO.

Civic society and symbols of community

The big organizational powers of other European countries in that period were defined by the monarchies and the church. The rise of the Dutch towns was special since they were organized by the people for the people in a model that has later been called the civic society. It is for that reason that the main buildings of the Dutch cities are not the palaces of its rulers but the modest and cost-aware town halls and assembly buildings of its citizens' government, the university, the theatres and botanic gardens.

Modest, efficient and functional

The church played a relatively minor role in the Dutch cities as religion was predominantly a modest Protestantism. Church buildings are present and visible in the Dutch towns but never a dominant force of organization. The fact that the Dutch were governed by the people for the people also explains why big urban elements that express absolute power of monarchy religion of government like avenues, squares and palaces are absent in the layout and the silhouette of the Dutch water towns. Looking at the Famous maps of cartographer Blaue, towns appear to be efficient walled trade harbors interspersed with small streets, little markets, singular houses and modest collective facilities. The most elusive signs of wealth were to be found on the facades of the rich merchant houses.

Stagnation

The Golden Age in the Netherlands lasted from around 1580 to 1700. Although Holland stayed a significant trade nation, after 1700 economic success suffered a period of decline. This is mainly attributed to the rise of other European nations as powers of international trade. England took over as the leading nation for trade and global influence. Netherlands fell under French domination and the Dutch towns did not develop significantly beyond the boundaries of the golden century and the era of the Dutch republic.

INDUSTRIALIZATION, MODERNIZATION AND SUBURBANIZATION

A slow start of industrialization
Many times the Dutch 19th century is described as a period of winter-sleep during which the Netherlands was still recovering from its boom in the golden century. The Dutch industrialization started relatively late and on a modest scale. However, many necessary modernizations took place, like introduction of rail networks and large numbers of canals and land winning projects. New focus came on urban development outside the city walls and around the train stations. By the end of the 19th century most Dutch cities had dismantled their defense walls and fortifications. New ideas about modern and hygienic city planning were applied resulting in water refresh management and sewer systems to provide healthier lives for the populations of most of the cities.

Rotterdam's expansion
What the golden century was to Amsterdam was the 19th century was to Rotterdam. The early effects of industrialization saw a significant expansion of the harbor of Rotterdam preparing it to become one of the world's biggest harbors' with strong rail and road connections to the European hinterland. The scale of the industrialization and the application of its evolving technologies became apparent on the south side of the river Maas in large docking facilities, machine-operated cranes and industrial storage of goods. At the same time Rotterdam became leading in more hygienic forms of urban development with the introduction of the Singel plan. Until early 20th century Dutch cities always had very unhealthy and stinky air caused by lack of sewers and open water refreshment. The Singels provided a water refresh system combined with green space for the crowded old city and new plots and parcels to provide its citizens with a mixed offer of high and low end residential and commercial facilities. Nowadays the Singels are to Rotterdam what the canals are to Amsterdam.

Introduction of systematic city planning
In the late 19th century the act of urban planning and design became an independent discipline and profession. Hitherto the planning of the city had been the domain of defense-specialists, water-engineers or incidentally architects. The construction of a city had a haphazard dimension to it. Mostly houses and commercial buildings were individually commissioned by the end-user. City construction came to a new level with the introduction of project-development whereby institutional commissioners started to develop larger construction projects or whole city blocks at a time. The new modus of development significantly sped up the urban growth.

Law on urban development and the rise of social housing
1901 saw the introduction of a long discussed introduction of the "woning-wet" (The law on urban development) since the housing for the masses severely lacked both quantity and quality. The major effect of this law was an enormous expansion of government funded social housing programs and a better regulation for the quality of residential buildings. Another effect was the legal obligation of towns to make binding legal plans for zoning and urban extensions. One can say that this law marked the transition from an organic to a more systematic planning of Dutch towns. It is here that the bottom up development of towns started to reverse into a more top down approach. The top down planning with the modernist way of separating areas for working, living, leisure and transportation interconnected by car roads would define much of the growth of Dutch towns in the 20th century.

Expansion
The period leading up to the Second World War (1940~1945) was defined by crisis economics and slow urban growth close to the historic downtowns. The period after the Second World War became the longest period of stability and economic growth in the history of global civilization. More than 2/3 of the Dutch buildings are built after the 2nd World War. The population increased from 9 million people in 1940 to 17 million today. The urban area in the Netherlands increased fourfold. This expansion becomes very clear if one looks at the growth maps of Dutch towns. One just needs to take a look at the historic boundaries of Amsterdam to understand the expansion after 1900.

Automobility and suburbanization
Probably the singular most influential factor in postwar planning was the explosive rise of car ownership and the introduction of a refined system of highways. Until the Second World War cities were always built with a certain compactness and proximity relationships towards the town centers and the areas of commerce and service. This came to an end as the car marked the beginning of the suburbanization and the spatial separation of living and working. During the large scale expansion of urban area in the Netherlands trade commerce and production moved out of the city centers and downtowns and was organized in industry and working districts outside the cities. It's important to note that with the increase of the built territory the density significantly decreased. Most of the residential growth was in low density developments containing singular dwellings with a garden.

Decline of historic centers and urban renewal

The suburbanization and departure of economic activities from the city centers led to significant decline of the town centers. During the seventies the problems of the abandoned and neglected neighborhoods, streets and buildings came to extreme urgency and the Dutch government started to initiate programs to reinforce and redefine the old downtowns and neighborhoods. This program was called "urban renewal" and it served to keep the old areas attractive and keep the original populations from moving to the suburbs and new towns. This social and economic program kept active for two decades until the cities became popular again as places for living and working.

Urban lifestyle becomes popular again

The last two decades have seen an enormous re-appreciation of urban lifestyle and living and working in town centers and downtowns, especially amongst young professionals and affluent citizens. The main cause is the rise of the service industry and the outsourcing of the manufacturing industries to abroad which asks for different aptitudes of the urban environments. This process of re-appreciation of urban conditions has been strongly endorsed by national government and backed by investment programs for city infrastructure, public space improvement and densification strategies. Processes of gentrifications and rising real estate prices prove to be strong upward forces behind the quality of the old urban neighborhoods. In many towns and cities the quality and quantity of buildings for all functions and activities within existing areas have been vastly increased.

Projects for city upgrading

The process of urban re-appreciation was underpinned by themed policy approach to upgrade and redevelop. Project typologies that have been dominantly applied are , for instance, 1. The train and public transportation hubs. 2. The upgrades of public space and facilities in the historic downtowns and 3. The revitalization and transformation of worn-out centrally located industry areas near the downtowns. Many Dutch towns today have excellent designed and engineered public transportation hubs, improved waterfronts, creative factories, and mixed-use office areas designated for comfortable living and working in the 21st century. Currently examples for ambitious urban projects are the kop van Zuid Waterfront development in Amsterdam and the Zuidas development in Amsterdam. Both are aimed at amplifying the connectivity providing global business hubs and creating new mixed-use high density areas for living and working.

Reversal of cities in a landscape to landscapes in a city

Not only have the Dutch cities developed beyond recognition, also the landscape in-between the cities has changed enormously. The amount of open landscape and agricultural area has decreased, and its functions have changed accordingly. The countryside and the open landscapes have increasingly been assigned as areas for leisure, recreation and ecology. In the recent past critics and thinkers of Dutch planning came to the collective realization that especially in the Randstad area transformation has occurred that the Dutch no longer live in a landscape with cities but in a large urban area with some open landscape. The core area of Holland is nowadays perceived as a continuous mix of towns and cities where the open landscape has defacto become a park structure. This new reality has led to the introduction of a new term, the urbanized landscape, referring to a reality where the classic boundaries of town and countryside no longer apply.

Leisure and agriculture in the open landscape

The intermingling of urban extensions towns and the open landscapes for agriculture result in a situation where the green areas are more and more accessible to its surrounding inhabitants and is increasingly transforming into a green context for the leisure economy. Many agricultural villages of the past in proximity to larger towns and cities have transformed into secluded suburbs with expansive houses, villas and facilities for leisure time like restaurants, coffee shops and other forms of entertainment.

Reading the urban agriculture landscape

Still there is an idealized view of the Dutch countryside and landscapes as the man-madeland of polders, inspiring geometry and natural beauty. That makes it easy to overlook that most of the organization of the Dutch landscape serves the goal of an economically efficient and reliable agriculture. Nowadays Netherlands Is the second agricultural exporter in the world and agricultural business has been heavily mechanized and has reached an industrial scale. Green open space that's not economically productive anymore is transformed into landscape parks. Not only have the cities and towns evolved dramatically, and the landscape as a complement has been transformed heavily as well.

CHALLENGES FOR THE CONTEMPORARY DUTCH TOWNS

Complete cities and feature towns

As laid out in the introduction, for the purpose of this book we distinguish complete cities and feature towns as complementary ingredients of urban development and urban regions. Complete towns are the larger urban agglomerations around big cities that have complete offers for working, living and recreation. The feature towns are the smaller urban cores that offer convenient living facilities for care, education and leisure while economically focusing on just a few vital features rather than a full spectrum of economic activity.

Dichotomy in a globalized economy

Without a doubt the main trend in Holland is the unprecedented popularity of "complete" cities. The opposite is true for the smaller towns in the more remote provinces. After the economic downturn of 2008 where the average Dutch housing price fell with almost 20 percent the complete cities have more then recovered. However, smaller towns in the provinces are still struggling to find the way up. Some regions remote to the complete cities even face depopulation and economic irrelevance. In the globalised economy of the 21st century there is an increasing divide of wealth and opportunity between larger cities and smaller towns. Although the divide in Holland is not insurmountable yet it surely presents a double challenge of managing sustainable growth in the large ones and keeping the small ones economically and socially vital and viable.

Urban Hotspots

Nice examples of inclusively developed Dutch cities are Amsterdam, Utrecht and Den Haag. They have identities still defined by historic remnants but they also seamlessly integrated all necessities for modern life. They offer a wide variety of residential styles and identities with good facilities and many different options for transportation. Usually they have one or more universities, one or more hospitals and wide varieties of shopping environments. Companies increasingly move here to be close to their allies, their competitors, the airport, the high skilled urban workforce and all the good amenities on offer. Individuals prefer urban life style of living in dense neighborhoods for high level of convenience and looking for opportunities. It's easier to finds jobs and clients and also easier to sustain existing friends and meet new ones. Nowadays also unexpected groups like families with children and affluent senior couples are attracted by urban lifestyles.

Challenges for the urban complete cities

Main challenge for the complete cities is to make sure they don't fall prey to their own success. Generally looking at Amsterdam as a successful complete city some major issues dominate the urban management.

1. As an ever more popular tourism destination large parts of the downtown are suffering increasing nuisance of tourism to the point where the local inhabitants feel the city becomes unlivable. The huge flux of tourism transforms the city into a leisure park threatening a well balanced offer of amenities, convenient transportation and affordability.

2. The popularity of urban living pushes up the housing prices to unprecedented levels. The un-affordability of housing leads to a class divide and drives many people, vital to the urban workforce out of the city to find better opportunities elsewhere.

3. To accommodate sustainable urban development it's vital to assign new areas and projects with the potential for mixed-use high density neighborhoods near service areas. The city's infrastructure needs to be upgraded and expanded to contain future growth.

4. The urban divide causes discontent amongst the inhabitants with their daily surroundings. Hence it's vital for the urban government to interactively use citizen's platforms to communicate, to discuss and properly inform processes of planning and development.

5. Address issues of economical and technical sustainability in the local streets and neighborhoods. Good examples are the individually built town houses and initiatives for green roofing, local water storage and energy production.

Vital towns

Netherlands counts many midsize and smaller towns in the proximity of the larger cities but also in the more remote areas and provinces. Generally one can observe that the vitality of towns is strongly related to the position in the urban networks and economic clusters as well as the presence of strong local assets like attractive living conditions, local heritage and landscape quality. In the present era of economic globalization it's the smaller towns outside the networks that become more vulnerable and will have more difficulties to stay attractive and socially and economically fit. Well educated people move to the larger cities and leave the older and less mobile people in the smaller towns.

Vital town challenges

In order for (feature) towns to stay vital and viable and to prevent decline, sets of interventions and policies can be defined. As usual this is not merely a matter of urban planning and design but more related to social and economic strategy.

1. Main challenge of feature towns is sustaining its strong features like industrial clusters of educational institutions. At the same time feature towns need to be on the lookout for new opportunities. Unlike complete cities new trends in economy and culture not automatically materialize.

2. One of the main challenges for the towns is to stay connected to the regional networks by car and public transportation. Nowadays revision of public transportation schedules or congested highways can be a threat to the towns' vitality.

3. Local towns need to protect the presence of amenities and services. The risk of young and affluent people leaving towns is the impendent collapse of facilities. Towns need to define a minimum level and protect it or organize it.

4. Preparing for shrinkage and depopulation is a realistic scenario for towns in regions with a negative demographic outlook in remote regions. Especially in the fringe of the Netherlands like provinces Zeeland, Limburg and Groningen active policies for shrinkage are being developed.

5. Protecting local features and local character is a must for those towns that rely on them for their unique selling points. A specific example is the towns along the Dutch coast that lose value by construction projects destroying natural beauty or livability.

FIVE FEATURE TOWN TYPOLOGIES

Categorization

In order to provide a good understanding of Dutch feature towns as complements for larger and more complete cities in the urban networks this book provides an overview of typical Dutch (feature) towns of varying scales and varying backgrounds. Developing a firm grip and a good understanding of the towns a categorization is needed that exceeds scale and position.

Variety in towns

In order to grasp the diversity and origin of the Dutch feature towns we have categorized them in easily understandable groups related to dominant parameters like historic development, economic and cultural shining points, and the infrastructure and landscape. This categorization is by no means scientific in its approach but serves to understand the variety of typologies:

1. History towns; Towns with historic tourism and leisure as the main industry

2. Water towns; Towns shaped by their unique relationship to the water system

3. Art towns; Towns with a renowned cultural contribution and art production

4. Product towns; Towns historically defined by the products of the local industry

5. Science towns: Towns living and breathing by their science community

Subjective positioning

This proposed categorization also integrates the way the towns are known to the public. For instance, Delft is famous for its product of Delft blue pottery nowadays, though it is equally famous for its university. As will become clear by the case studies, character and identities of towns can change overtime. Some Dutch university towns started out as centers for trade and defense, other towns have an origin as centers of agricultural trade but we reborn in later areas as centers of art culture and leisure. This book captures identities as they are perceived today. Many towns can be simultaneously defined as a water town and a product town and a history town based on how you look at them. However, to create a proper understanding of the Dutch towns we think the chosen categorization is didactically very useful.

The history town

The history town is used to describe the category of town that somewhere along the line got stuck in their development and have been safeguarded against the modernization during the 19th and 20th century. Often these towns have remote locations and were dependent on specific assets or features like trade, local production or defense roles that disappeared over time. Nowadays these towns have become frozen history by keeping the same shape and layout and buildings as were built by long gone ancestors. This type of towns has found second or third lives as tourist attractions that rejuvenate the local business and local economies.

The water town

The river, the waterways, and the water system are relevant to all Dutch cities. Some towns have an exemplary relation to the water emphasizing aspects of trade, defense, transportation, water management and community development. The nature of this relation heavily influences the urban design in the way of structure, public space and how individual buildings are positioned towards the water. In many Dutch towns the historically delivered land-water arrangement gave rise to special types of communities and new perspectives on leisure and value creation.

The art town
The art town is used to describe towns where people retreat to focus on cultural production and alternative lifestyles. These towns, usually in the proximity of larger and more dynamic towns, have produced individuals and movements that are named after the towns. The reasons for attracting specific groups are related to the local charms of architecture, landscapes like forests and coastal dunes and convenient infrastructures.

The product town
It has often happened that local products like beer, cheese, porcelain of technological artifacts precede the name and fame of towns. The town name becomes a brand name for specific products. The physical contexts for this production like buildings, trade centers, infrastructures or marketplaces usually give these towns distinct characteristics and become attractions for local tourism. Although new products and services may have surpassed the old ones, the identity lives on.

The science town
In the 21th century innovation and knowledge production have been positioned at the heart of the global economy. Towns with academic research are becoming epicenters of economic transitions but at the same time they have long historic roots. Science towns organize themselves to accommodate their students, researchers and research institutions. At the same time they aim at becoming attractive and comfortable living environments where life has more to offer than science.

HISTORY TOWNS

Veere Zeeland
Water sports paradise

Today if you bike along the famous Dutch Delta works and take a turn at at the Veerse Meer with all it leisure boats you will be impressed arriving within the citywalls of Veere. Veere is a little town with a long and telling history in the south of the Netherlands on the shoreline of the North Sea. The immaculate preservation of this famous little Dutch town and its building makes it easy to imagine how people lived a long time ago. The story of Veere reads like a thriller; humble origins, war, success in business, but also disasters. Veere had to re-invent itself 4 times to arrive where it is today.

Once a wealthy international sea harbor, Veere is now at the middle of a large water sports recreation zone around the Veerse Meer. The Veerse Meer, a brackish lake, has a length of 22 kilometers, counts 13 islands and includes six yachting harbors totaling over 5000 leisure boat docks. The lake is a hotspot for national and international water recreation tourism. The water sports economy is hugely amplified by the wonderful rural landscapes of the Zeeland province and the proximity of the North Sea beaches with many tourism and recreation towns in the direct vicinity. Although the water recreation of privately owned leisure boats can be considered a post industrial 20th century phenomena, the history of Veere as a little defense town goes almost 700 years back in time. Veere has had many different lifecycles to get to where it is today.

Walking around Veere, one is amazed at the immaculate historical state of the proud little town. Entering from the landside through the gate in the former defense works, one arrives at the long linear harbor called "Kaai" (the quay). Parallel to the Kaai 100's of yachts are docked all year long and make for a visual pleasure of boats with historic architecture as décor. From the "Kaai" a few streets lead inside the town and closest to the open water a small street widen to the central area of the town called "markt" (the market). The market is surrounded with powerful historic buildings bearing an intensity that immediately signifies the stronghold Veere once has been. All year round Veere is bustling with dedicated tourists strolling around, gaping at this town that enables them to imagine living 400 years ago. With more than 150 state registered monuments within the cities fortifications in a comprehensive grid of little streets there is a lot to see and experience.

Going back and peeling of the layers of Veere's History, one can decipher a number of stages or phases that left their marks on the town as it is today. Veere started out as a little fishing village in the 13th or 14th century. Given the relatively protected position towards the open sea, it developed into a transit sea harbor and due to its geography and trade connections to other Dutch towns it grew in importance. In 1541 Veere became the official liaison town for trade between the Dutch and the Scottish. The Scottish would bring in wool and it would be traded for Dutch roof tiles, sheets, greenery and weaponry. In this period Veere became a wealthy town thanks to the trade and a significant Scottish population. Another important role to be played by Veere was defined by its strategic military position on the south side of the Dutch republic in between Rotterdam and Antwerpen. During the hectic periods of the 80-year war with the Spanish empire that lasted from 1568-1648

Veere became a stronghold in the coalition of the Dutch republic. These days Veere was counting 3000 inhabitants (double of nowadays) and of these 300 were Scottish trade people and administrators. Around 1800 the the Scottish influence and the prosperity it brought along came to an end under the French rule in the Bataafse republiek. From a heavyweight trade and military town Veere returned to the obsolescence of fishermen's village again. Its internal harbor was home to a meager fleet of thirty fishing boats for many years to come.

The town as we can visit it today was formed around 1600. All the important buildings and the town's fortifications date back to this historic episode. The fortifications surrounding the town and harbor have a double objective of protecting Veere from both its enemies and the sea. The fortifications grew in 3 stages. A rudimentary wall for flood protection was completed around 1500 after series of severe flooding. The second phase of fortification construction had a more military approach as Veere got to be an important stronghold and improved defense security became a necessity. The final phase in the fortification was realized during the Napoleontic area in the beginning of the 19th century when Napoleon decided Veere's fortifications should be extended with a second line of defense. Currently the once military fortification can be seen as a big piece of land art that provides multi angular perspectives on the silhouette of Veere's impressive and heroic legacy, but also on its surrounding water landscapes.

Veere counts a number of top monuments. The Scottish left a row of townhouses on the quay of Veere as remnants of the Scottish-Dutch trade relations. The houses date back to 1540 when they were initiated by the dignitaries of the Scottish trade community. Nowadays the houses serve as Museum Veere. The Town Hall of Veere is a silent witness of Dutch prowess and cunning in the 16 century. The construction started in 1474 at the height of Veere commercial and administrative powers. The building is constructed in a Gothic style and the impeccable geometric order of its main façade integrates 7 statues of Veere's most important 15th century administrators and their family members. One of the most exquisite features is a carillon that can still be played today. The old town hall nowadays serves as a special event facility and is favorable as a wedding venue. At the same time it operates as an exhibition on Veere's History with many "sur place" artifacts.

The largest building in the town is the main cathedral that started construction in 1479 in the hey-days of Veere. The cathedral has been extended and changed in outlook, function and use in successive historic episodes. The architectural result is an amalgamation of styles and spaces that reads like a short history of Veere itself. From the start the church had three different religious groups, each using a partition of space around the nave. During the Spanish wars most of the church's treasures disappeared. In the 19th century the church mainly served as a hospital and some parts were demolished and sold off as construction material. Nowadays the revised building serves as a concert venue and cultural event center.

One would think that Veere had been through enough experiences. However, its most profound trajectory changing event is from more recent times. The protection from the water, both the sea and the water coming from the European hinterlands through the dutch estuary was badly managed and maintained since the late 19th century. In the early 20th century the first and second world wars and the intermediary economic crisis in the 1930's, led to an accumulation of risks by neglect. The flood protection for Veere and for the province of Zeeland as a whole started showing many weaknesses. On a disastrous day in February 1953 a super-springtide proved to almighty for the water-defense systems. Dikes started braking down one after another and as a result 1000's of square kilometers of Dutch land inundated, with the water swallowing everything that came on its way. All normal life had been suspended and it took a few days for the Dutch government to understand the scale of the event, given that all means of communication had been disabled by the flood. A large (inter) national rescue program was organized and it took a long while before significant relief was realized. The infamous February 1953 disaster took about 1500 people's lives in the South of the Netherlands and had a huge impact on the future development of the dutch coastline and water management programs.

Until 1953 spring tide disaster, Veere was still directly connected with the North Sea. The seacoast along the islands of Zeeland was hundreds of kilometers long, giving many little towns also more inland direct connections to the sea. However evaluating the springtide disaster and formulating preventive strategies the Dutch government came up with the ambitious plan to drastically reduce the seashore coastal length of the Zeeuwse Islands by building a large defense system of dams along the coast that would bear the famous name of "Delta Works". The delta works regulate the water flowing out from the rivers into the sea and protects the land from springtides. The consequences of this strategy for the province of Zeeland were enormous. Many little towns and villages lost their direct access to the sea. Towns like Veere lost their sea harbor and their connection to the open sea which led to the decline of the local fishing business. The ecological consequences, however, include the disappearance of tides, another salt/sweet water balance but also a more controlled water system to guarantee water safety and ecological diversity and to provide for leisure boating and water sports. The position of Veere as a sea harbor at the Veerse gat with salt water transformed into Veere aan het Veerse Meer with brackish water.

The delta works provided a strong impulse for the development of Zeeland as a tourist and leisure region. Along the road that connects all the dams along the sealine many easily accessible resorts have sprung up. In the last 30 years Veere has found new economic activity as the center of 6 leisure harbors in the Veerse Meer surrounding Veere. The town has become a hotspot for both Dutch water sport tourism and Germans looking for a holiday near the Zeeland beach in a quiet setting.

Many tourists from the surrounding holiday villages and resorts like Vrouwenpolder Kamperland, Ruiterpsaat, etc. will visit Veere for a walk, a coffee or a meal. Nowadays Veere receives tens of thousands of tourists that arrive by boat, by bike and by car.

For designers and planners it's very relevant to note that this magnificent recreation landscape of a large recreational lake with a historic tourism destination is not just a product of design. On the contrary, the whole setup around Veere is more like a delivery of history that is put to its best use and creates an experience that is deeper and more emotionally fulfilling than a jus out of the box design could ever be. Also it's important to realize that the magnificent fortifications, the petite harbor quay, picturesque houses, the small streets and squares leading to the major churches and the town hall are silent witnesses of big episodes in Dutch history. One of the most impressive features of Veere and its surroundings is its resilience. Veere had to reinvent itself 4 times completely in terms of economy administration and access. From a trade and defense industry to administration and war, to become a fisher town again and finally reinvent itself as a center piece of the water sport recreation industry, thanks to a large natural disaster. Veere and its surroundings teach us a lesson. To see a historic destiny one needs to look way ahead of the current state of affairs.

Volendam

Volendam is a famous tourism town in North Holland Province, with the coverage of 16 square kilometers and a population of about 22 thousand. Previously as a fishing village, Volendam encountered the fishing industry decrease after the dam was built, but still kept the old style. Now with tourism as the main industry, Volendam received millions of visitors from the world each year.

Volendam is established and developed from the ancient city Edam. Edam, which obtained the ruling rights in 1357, is a famous ancient city and harbor city in Holland. To expand the city development in the harbor, Edam opened a new port at the island of 3km^2 in the northeast direction. To separate the newly opened area with the South Sea, a dam was built to separate the old harbor from Edam city and the South Sea. The closed dam Volle is the origin of the name of Volendam. After the completion of the dam, the fishermen moved here to facilitate fishing in the South Sea. Since then, a famous fishing village was formed quietly.

After several centuries, Volendam grew and developed. There were about 2000 fishing boats in 1800 and during each fishing time there were hundreds of fishing fleets. At that time, the fishery played a very important role in the economy of Holland. Because of the opening of the North Sea Canal, fishing boats in the South Sea began to go fishing in the North Sea.

Although Volendam is close to Edam, it is surrounded by water and can be reached only by water. This isolated way of living makes residents in Volendam maintain their unique lifestyle without being disturbed by the outside world. They do not contact with the outside world, until after 1860 people outside moved in. It is also this physical isolation living way and Volendam people trying to become the best of the best in nature that enables them to well preserve their own unique life habits, dialects, folk customs and clothing.

Men help each other during fishing, which is the reflection of friendly and cheerful characteristics of Volendam people. Women always make the traditional costumes while waiting for men back from fishing. When the long-awaited fishing fleet came back, the whole village was bustling. People smoked the live eel immediately, sang and danced to celebrate the harvest. Today, a large number of famous Holland folk singers may also originate from this tradition.

The opening of the North Sea canal and the establishment of Ijmuiden fishing port makes the fishery gradually prosperous there. The continuous flood brings great disaster to the area around the South Sea, which makes the government build the dam to completely solve the flood problem and alleviate the problem of land shortage through reclamation. Although this amazing plan was suspended due to Volendam fishermen's strong opposition and protest, the flood disaster in 1916 finally made the government determine to reclaim land from the sea. After the dam completion in 1932, the Volendam fishery was hit greatly. Many fishermen did not want to leave the hometown and gradually changed their profession.

Located along the dam, Volendam relies on the fishing industry. The construction and planning of the town is around the harbor. Before the construction of the dam, Volendam is a true fishing village. After the dam establishment, traditional fishery gradually disappears and Volendam successfully transforms into a fishing village relying on tourism, and even known as the world's most beautiful village. The key lies in its unique culture charm, which is the internal characteristic to form Volendam feature. The unique Volendam culture that consists with its unique folk custom, architecture, traditional clothing and way of living attract people from all over the world to experience the harmonious life in this peaceful town.

Since the beginning of 19th century, people in Volendam have been very loyal to their clothing, which is beautiful and elegant with fine workmanship. Along with the time, they gradually form their own style of clothing and till 60s in the last century, people always dressed traditional costumes. Nowadays, the world famous Volendam dress is the style at that time without changing.

Volendam clothing, especially women's clothing, is world famous, which becomes the Dutch symbol in the national publicity film together with the windmill, cheese and wooden shoes. The reason that Volendam clothing can obtain such a high international reputation is closely related with many world famous artists who came here around the 19th Century. In 1873 the French traveler and historian Henry Havard mentioned Volendam for the first time in the tourism biography. Later British famous painter George Clausen presented Volendam costumes in the painting and then the photographers fell over each other to come to Volendam. Figures of strong fishermen, women dressed in colorful national costumes and children playing in front of the beautiful oil painting houses appear in various magazines. Chased after by many famous painters, the town became a place for all painters. Spaander Hotel was the favorite place for those artists at that time. Now the hotel has changed the name of Spaander Art Hotel, where artworks of these famous artists are exhibited and most are used as the hotel fees. A large number of well-known artists further enhance the visibility of Volendam fishing village.

Let's introduce the features of Volendam dressing which become beautiful scenery in Volendam tourism and a national symbol of Holland.

Women's clothing starts from the hat, which is a very precious lace cap. To make the cap in a special shape is like an art, needing careful work. The most prominent thing is the three-circle red coral bead necklace with golden lock, a variety of embroidery apron and beautiful blue-white tie with tassels. The tie is the classic part in Volendam women's dressing. Outside the apron is the black cotton or wool blouse with half sleeve. The tops are closed with hidden hook loops, and then leveled by a blue wool line. The down long dress has diverse colors. The first time to wear red striped or colorful dress must be in the wedding and then in holidays. Wear grey black blue striped dress on Sundays. Wear black ruffle dress while needling and pure blue dress whiling working. Outside the dress there is an apron. The ordinary apron is mostly made of blue cotton striped cloth and that in holidays is made of black wool cloth with embroidery in the waist part, which matches with the embroidery in the upper body part and makes the dress look more beautiful.

Men wear the black collar with beautiful embroidery. Two golden buttons on the collar means adult and silver one for teenage boys. The shirt is normally with red or white stripes, sleeveless, double breasted. The outerwear is black wool with six buttons. Only the following three buttons are fastened as well as the cuffs. There are silver chains or other accessories on both sides of the collar. All tops are into pants. The trousers are usually black velvet, which are very fat. There is a square cover in the front of the waist with two silver buttons on the waistband. They often wear black silk eel like a collar with the end tied into the coat collar, which is a classic part for men clothes in Volendam.

It is indispensable to see a blue collar on a girl's neck, which is the highlight of girl dress in Volendam. Today when walking across the streets and lanes of Volendam, you can see the local people dressed in traditional costumes, coming out from the well planned wooden houses with green facades and greeting each other in the dialect. Elegance is presented from this quiet picture. People visiting here can experience the uniqueness in Volendam. The unique and profound historical and cultural origin is deeply rooted into the blood of the people in Volendam, which is their pride.

Giethoorn
A village from the movies

Is it possible a town becomes immortal after being used as a movie set? Yes it is! Giethoorn became immortal in the Dutch movie called "fanfare" telling a story of competing orchestras playing on boats. Nowadays Giethoorn is an international tourism magnet located next to the national park of Weeribben. A former agricultural community was placed within a remote and unique historic landscape. The mosaic of water and land wasn't suited for roads and resulted in a system of farms, houses and facilities on little islands interconnected by boats via small waterways.

Modernization and progress don't spread equally over a country. Far away from the large cities, the engines of progress, there are voids in the landscape where time and development come to a standstill. In these voids the habits and rituals of the past continue and the landscape doesn't change. This especially applies to the landscape around Giethoorn where the nearby river called IJsel flows into the IJselmeer. Giethoorn was part of a larger agricultural region of little villages and small-scale activities, of which none grew very hard or developed beyond a certain level. Giethoorn is one of those rare communities where time came to a standstill and where one is able to get a exclusive peek into the past.

Long time ago Giethoorn was a remote landscape of swamps, forests and peat bogs. The population density was very low and concentrated in few isolated settlements and villages. In the early times peat harvesting was one of main sources of breadwinning. This went on for ages; the peat would be loaded onto little ships and transported to be sold on the markets in the nearby towns. Relentless exploitation of peat landscapes by removing the soil led to strong deformation of the landscape and it turned into an organic mosaic of land and water. The earliest settlements moved along with the patterns of the peat harvesting. Not until the 17th century did the village of Giethoorn get a more permanent shape and position. At a certain moment the locals reached a limit and the winning of peat became gradually impossible. Around 1750 the local economy diversified and winning of peat was extended with cattle breeding, fishing and cane-cutting for roofing and other applications. Many people living in the region simultaneously had three professions; farmer, fisherman and cane-cutter in order to be able to make money as seasons changed and other opportunities were scarce.

Given the fact that the land was continuously interspersed with water, roads on land, and therefore transportation on land, were not possible. The internal transportation in Giethoorn was organized by small boats that could carry anything, from the daily distribution of fresh breads to the incidental funerals of the village people. The outbound transportation consisted mostly of agricultural products and products related to the growing of canes and reeds. Exports to the markets in the nearby towns like Meppel and Steenwijk were organized on small boats that could easily navigate on smaller waterways. Until the early 20th century the village was disconnected from any road systems on the land and even today Giethoorn is not accessible by car.
It was not until the 1950's that Giethoorn was rescued from oblivion by one of the most successful movies in Dutch history called "fanfare". The Director of the movie Bert Haanstra was the hero of the early Dutch cinema and his movies won awards at the Cannes film festival and one of his later works, a documentary about the production of glass even won an Oscar. The movie "Fanfare" depicts two bitterly competing fanfare flourish bands that split the village community when they are to battle each other at a local music festival. Both bands practice in the barns of Giethoorn and move around on boats depicting the village life. On the day of the festival the bands end up simultaneously on the stage and their music miraculously melts together winning them the trophy and reuniting the village. Bert Haanstra, the director managed to reduce the essence of Dutch character and Dutch culture to a few recognizable personalities and a very distinct village background with some humorous twists. Given the fact that Haanstra was born in the province of Overijsel and knew the landscape very well it is hardly surprising Haanstra came up with this exquisite village as décor for his movie. Giethoorn had hitherto been virtually unknown. The movie proved to be a big hit and millions of Dutch people saw the good-natured comedy with the sharp definition of the Dutch culture in the cinema. Overnight Giethoorn became a popular tourism destination and tourism became the most important source of income for the villagers. The village got a kind of celebrity status and with the disappearance of the local agriculture many well to do people choose this cool place to buy an abandoned farm as their second house for leisure time.
With the (re)discovery of the village also the surrounding landscape became a hotspot and was assigned as a national park going under the name of "Weerribben"

and "het Giethoornse meer". This natural reservation has a size nowadays of around 35 km² and is unique even in Europe for a natural swamp-style landscape of this size and quality. The landscape with its open horizons, reflections of light on the water, and many rare species of flora and fauna is ideal for biking, boating and hiking and experiencing nature. For many Dutch people the "Weerribben" is a familiar destination for a day out or a holiday in their own country. The "Weerribben" and its surroundings have repeatedly won the award for the most beautiful spot in the Netherlands.

In the fifties and sixties of the last century many plans have been made to on local and provincial level reorganize and rationalize the landscape of Giethoorn to provide better opportunities for local agriculture, industry and transportation. All those major efforts however have been vetoed by the national government. In those days the national council for landscape and culture deemed the region around Giethoorn too precious and too unique to be brutally modernized. All of this took place even before the village had become immortal by the movie. As a result the village and its surrounding landscapes kept its historical identity and modernization only took place with respect for historic and cultural values. Nowadays Giethoorn is a protected heritage site (1981) where construction and reconstruction are severely limited and subjected to strict criteria to protect the identity and the landscape and keep the memory alive. This decision of the national government contributed many years later to the blossoming of the tourism and leisure economy.

The village still has its historic layout of a longitudinal shape subdivided in three cascading parts called "Noordeinde", "Middenbuurt" and "Zuideinde". The three parts are connected by the village canal that serves for the main transportation the typical village boats. The typical farm houses are called "Bump farms" because of the typical roof shape that starts with small residential quarters and extends into bigger barns under the same roof. The farm houses are separated by small ditches that can be crossed by small privately owned bridges, high enough to let the low boats pass underneath. The village counts almost 200 of these small bridges. As the village is a rare relict, so is its boating culture. Various types of boats have historically developed to suit different functions of transportations with passing under relatively low bridges as design specifications. In sliding scales there are the "Gieterse punter" for transporting indivuals, the "Gieterse volt" for distributing small loads and the largest one, the "Gieterse punter", designed for transportation of products and foods to nearby market towns. All of the boats are very flat allowing them to pass under relatively low bridges. In the smaller waters they would be rowed or punted and in larger waters these boats were able to sail. Nowadays the same boats have been equipped with diverse motorized solutions, mainly electric to keep the peace and the quiet of the village.

Due course of time a strong tourism and leisure infrastructure has developed as a steady stream of tourism has found its way to "Giethoorn". Tourism has become the main pillar of the local economy with little hotels, restaurants and rental services for boat trips as core activity. It is a little miracle that the small magic village is an especially popular destination with tourists from Asia. Nowadays more that 200.000 tourists from China come to see this miraculous Dutch water village as a Venice of the North. On their travel itineraries "Giethoorn" is put along destinations like Paris, Amsterdam and London. As modernization is omnipresent in the globalized cities around the world in the 21st century, Giethoorn is still a rare exception as a little space that escaped modernization and provides an oasis where one can experience an era long gone by.

WATER TOWNS

Vinkeveen - Ronde Venen
Watergarden for the rich

The polder land south of Amsterdam is home to a poly-nuclear little town called Ronde Venen. The name of this town is based on the breathtaking geometry of the landscape of a circular water structure with green wedge shaped meadow parcels and a number of little villages. The area once was a peat colony of Amsterdam where the peaty-land was stripped and shippe, leaving behind irregular arrangements of land and water. Nowadays the peat winning areas of the past are celebrated water sports lakes and exclusive islands built with villas for the rich and famous of the Amsterdam region.

The formation of the Dutch man-made landscape has taken centuries to arrive at its current form. Man-induced action for agricultural production and water management resulted in the transition of natural-landscape into culture-landscape. An eminent example of this culture-landscape is the area south of Amsterdam now called Ronde Venen. The shapes and forms of parcels and waterways going back as early as 2000 year ago still define the landscape. For Ronde Venen round curved rivers and water flows with names as "de Kromme Mydrecht" and "de Amstel" are very characteristic. The shapes of these rivers are formed by melting waters of the glacial periods when Netherlands was covered under large ice structures. These round water structures and the peaty land gave the area its name (Ronde Venen).

For agricultural purposes the land was divided in many similar parcels with dimensions based on keeping certain amounts of cattle related with the scales of early farming. In order to keep the land dry it was essential for each plot to have longitudinal ditches perpendicular to the curved rivers and the water flows. Hence in the 13th and 14th century a pattern of wedge shaped parcels arose, each with a wide front to the river and a smaller front toward the middle of the land. A most remarkable landscape geometry was shaped as a consequence of the curved ancient water flows. Over the centuries, the peaty lands sunk lower and lower as a consequence of the agricultural activities and it became more and more difficult to keep the area dry and fit for economic purpose. Hence new canals and water flows were created and early civil engineers created advanced water systems to discharge the water towards the sea.

Another powerful force in the formation of the landscape was the peat harvesting. In the Netherlands the peat harvesting in-between 1550 and 1950 transformed more than 100,000 ha of natural landscapes. Already in the 14th and 15th century peat from the area around Ronde Venen was harvested from the land and sold to the larger cities like Amsterdam as fuel of all sorts. The peat distribution to all the homes in the towns was a big business that required canal boats, turf markets in the towns, but fore-mostly peat harvesters in the large peat areas outside the towns. For the local people of Vinkeveen and Ronde Venen the peat harvesting was a temporary solution to generate income. The landscape however suffered long term consequences by all the disappearing land and all the water transportation facilities that were necessary to get the peat to the cities.

By harvesting the peat the water surface became larger and the larger water mass was a danger to the dry lands, that also sunk due to agricultural activity. More and more interventions to keep the remaining land dry were necessary. The application of windmills served to discharge water from very early on and remnants of all kinds of windmills can still be found all around Ronde Venen. One can imagine that one of the important political themes in the governing bodies of towns like vinkeveen has always been to keep the land dry enough to maintain economic production to feed its citizens. Nowadays the management of water, drinking water production and water safety is organized in "waterschappen", regional water authorities that form an important subsystem of Dutch government and administration.

The geographic area of Ronde Venen is home to a network of villages that have administratively joined since the 1990's to form a relatively small polynuclear town. The merging of villages into little towns is a process that has been applied regularly to make the administration of regions more effective and to keep the costs of government low. The names of these merged cores are Abcoude, Meijdrecht, Baambrugge, Wilnis en Vinkeveen. Meijdrecht being the largest core is the main settlement with a larger amount of industry and inhabitants. The eastern boundary of Ronde venen is the national Highway A2, the most important economic artery of the Netherlands. The proximity of both Amsterdam and Utrecht by highway makes for a diverse and regionally oriented focus of businesses and inhabitants. Ronde Venen is an integrated part of the urban network in the middle of the Netherlands that is attractive by providing urban opportunities, yet offering rural and landscape qualities for living, working and especially recreation.

The most illustrious part of Ronde Venen is Vinkeveen, in one of zones where the peat extraction had the largest influence on the landscape and left large lakes and countless little islands that were economically useless for long periods of time. However, with the rise of the leisure industry and the individual wealth growth of the Dutch people the lakes transformed into water-sports areas where many families own a pleasure boat or go for surfing, diving and hiking. The little islands were transformed into private paradises with vacation houses. In very prosperous period of th 1990's on many of the larger islands large luxury villas were built. With its brand of water sports and exclusive villas Vinkeveen has become a famous name in the Netherlands. Nationwide Vinkeveen is unique as a habitat offering the combination of a central location with a good access combined with breathtaking landscape and exclusive villas on water parcels. Many successful Dutch people have lived in Vinkeveen, and moving to Vinkeveen became synonymous with a very successful career with massive financial benefits. Maybe the most famous inhabitant of Vinkeveen was the legendary Dutch football player Johan Cruiijf who lived there for 20 years and was one of the first to have a large villa built along the lakes in the 1970's.

The Vinkeveense plassen, as it is called in Holland, is a cluster of three lakes with a total surface of around 10 km^2. The lakes are praised for having the cleanest surface water in the Netherlands. The shape of the lakes and the organic position of the 100's of small islands and strips of land are the result of ages of peat harvesting in a landscape that historically made wedge-shaped parcels. In the last century the area was also used for sand extraction and at some areas the water has a considerable depth of about 53 meters. On the whole it forms a land and water framework with a magic spatial structure that organically developed over time. It's safe to say that a landscape was generated that no designer would ever have envisioned. It is one of those magnificent examples where landscapes of the past have found new uses in new times. Experiencing the area from a boat it shows a large biodiversity of shrubs, reeds, trees birds, and fishes. The area is very popular for diverse water sports like sailing, surfing, boating and diving. Around the waters there are a number of leisure- marinas and camping facilities. The numerous islands in the lakes have mostly private uses with recreation pavilions that are only accessible by boat from the nearby marinas. Some islands in the middle of the lakes have a service function for small commerce and sanitary facilities.

One of the main axes in the area, a division in-between the lakes is a long strip of land with a small road connecting Vinkeveen and Baambrugge. This landscape element is called "de Baambrugse Zuwe" and is famous for its villa parcels facing the lakes along the road. The villas have parcel sizes in-between 2000~5000m^2. The architectural styles form an amalgam with classic villas, modern villas and anything in between. All the villas have undisturbed views on the water and all of the villas also have private jetties for one or more yachts to enjoy the lakes by boat. Although there have always been planning restrictions and guidelines in the area regarding volume height and the profile of the street, the whole spatial organization of the villas islands and lakes are a product of organic development that took centuries to evolve. One of the jokes about Vinkeveen is that is has become a hangout for the happy few, or maybe even a show off zone for people who like to show they have too much money. There is no doubt, however, the landscape and the lakes are a joy to all.

One can also say this area shows the irony of history; the formation of the landscape by agricultural activity and the periods of peat harvesting were characterized by centuries of hard work and low pay. In the new era the same landscapes resulting from these hard and tough activities have now become the habitat of the rich and the famous urban people looking for relaxing leisure time.

Dordrecht
Star from the past

Dordrecht is one of the oldest Dutch water towns located at the island of Dordrecht and surrounded by three mighty Dutch rivers "De Noord, de Merwede en de Oude Maas". Historically Dordrecht played an important role in the trade between the southern and the northern Netherlands. As the trade and logistics moved in the direction of the sea to the Rotterdam harbor, Dordrecht gradually lost the power of its heydays but along its rivers still an impressive industry of shipbuilding production and transportation technology can be found. Nowadays Dordrecht is a comfortable Province town with a rich historic past.

Dordrecht is founded at one of the currently busiest three river point in Europe. It's the place where the Merwede splits in the Noord and the Oude Maas, both going towards Rotterdam and the North Sea. The story of Dordrecht is the story of the river, the river boats, the trade and transportation economy and a lost battle of becoming the biggest and most important harbor of the Netherlands. Even nowadays, 400 years after the heydays of Dordrecht the feel of wealth and power on the historic quays along the Oude Maas, the Wijnhaven, and the Riedijks haven will fill people with awe. Today it is also hard to imagine that Dordrecht, a well connected part of the Rotterdam agglomeration, has for most of its existence been a town on an island only accessible by boats and ferries. It was only in 1936 that Dordrecht got to have a land connection by a long bridge allowing for a 24hrs-a-day access to the town. The longstanding joke has been that the people of Dordrecht think differently because they have an island mentality by living on an island for many generations.

From the very early times Dordrecht was strategically located on a crossing point of trade routes, from Antwerp to Amsterdam and from Brabant to Rotterdam. In these periods the roads on land were unreliable and closed in winter. Transportation on wheels hadn't been invented so larger quantities goods and products were naturally moved by boats over waterways. The city of Dordrecht was a transit harbor where the ships with goods would sail in the town. The goods would be unloaded and stored in warehouses and sold off or traded the days after in one of the many local and specialized markets. The integration of trade, waterways, harbor storage and a specialized workforce was relatively very advanced, compared to its peers in Holland and other parts of Europe. The city on the meeting point of three rivers optimized for trade on water could grow very fast and become very powerful. In the 14th century Dordrecht was handed transportation- taxation rights by grace of the ancient government of Holland called the Dutch County. This special status as a "stapel plaats" or as special trade zone for trans-shipment allowed Dordrecht to collect tax over each transition of goods that were unloaded and traded again or simply fare through the town. This way the city gained capital needed for its expansion and investments in local infrastructure, local government and local welfare. The special status allowed the merchants of Dordrecht to live well and they could build wonderful mansions for themselves. Many of the public and private architectures of this era still embodies wealth today.

The historic town center is relatively modest in size with an area of 2 km^2, yet it still offers glimpses in the maze of medieval streets, alleys, Quays and harbors. The main organization is defined by two concentric waterways bringing the ships into the town. A third concentric line was defined by the former urban defense works. The layout of the canals and harbors tries to create an efficient harbor layout within a town surrounded by a wall. Since the wall was extremely expensive to build and the land needed to be won from the water, the town needed to be as compact and efficient as possible. As is commonly known, a circle combines a minimal perimeter with a maximum surface, hence the shape of the town is circular and located on the corner of Merwede and Oude Maas. Looking at old maps one can see that all the buildings efficiently positioned along the waterlines and the more extensive functions like basic agriculture and production are positioned outside the city walls.

One can grasp that the functional organization of the town was based on the types of products that were trades and the kind of space and facilities they need. The old names of streets quays and harbors provide an insight in the way the different types of traded were distributed over the city. The wine harbor is next to the wine market. The meat market is next to the pig harbor, and the market for construction materials is next to the Kalk haven where the ships with construction materials were docked and unloaded. Given the rough nature of the material, they were positioned on the outside. This also applies to the Taan the fluid that was used to make wood and sails of ships sustainable and weatherproof, the Taan market was next to one of the main ship entrances of the city. The wine harbor was next to the wine market and the wine street, so one can imagine the functional zoning of the town by products and industries. Basically Dordrecht and many of the other Dutch water towns were organized market places zoned by product types and industries with a very high level of integration of waterways, unloading docks, marketplaces storage facilities, and of course life facilities.

On the same map one can decipher more interesting aspects of life in this classic water trade town. The gates in the town all serve specific purposes. There is a gate for dairy products, there is a gate for transportation of goods in bags, but also there are gates for the local military and even a special gate for the provost of the town not to be used by other people. so the access from the water and from land was strictly organized and regulated. Looking at the map, it is significant to see in the relatively small town of Dordrecht there were more than 10 churches of different signatures and different religious beliefs. It shows that the church and religion played a central role in the daily life and organization of a community. Also looking at the map it becomes clear that the central area of Dordrecht was home to many social institutions like elderly care for men and women separated, for sick people and for the local government. All in all the map shows a high level of organization that was put in place to make sure the town was run efficiently and safe and beneficial to its inhabitants. One can be assured that the degree of organization that is shown by this map was also present in all the other Dutch towns of the same era in which the network of Dutch towns formed the Dutch republic in the 16 and seventeenth century.

It is not only good fortune that propelled Dordrecht's development as two big disasters came to Dordrecht and heavily influenced its history. In 1421 the big St Elisabeth's flood came to reshuffle the map of the Netherlands. Dordrecht was cut off from the Hinterland and became permanently a town on an island. The distribution area for its trade that could be reached by land was seriously reduced and so were its trade profits. In 1457 a devastating fire blazed the city into ashes and meant a severe setback in its development. Still in this period Dordrecht was one of Holland's most advanced and powerful cities. In the gathering of cities that governed Holland Dordrecht held the first chair which meant they also need to deliver the main administrator (de raads-pensionaris) of Holland. Another consequence of its first position was a leading role in religious matters. In 1619 Dordrecht became the meeting place for one of the most important religious conferences in Dutch history called the Dordtse Synode. This conference lasted over 200 days and was of big consequences for the role and interpretation of religion. Representatives of all the Netherlands and abroad came to discuss the righteous interpretation of the bible and decided to unify behind one proper Dutch translation called the "Staten Bijbel". This edition of the bible is still serving today. The trade benefits that made Dordrecht one of the richest and largest town of the Dutch republic slowly came to an end as larger cities like Rotterdam and Amsterdam started to become more and more important. During the Golden century Dordrecht couldn't keep up with the success of it neighbors that were better positioned and better suited for trade overseas. Dordrecht came to a slow decline and eventually turned into a poor isolated town on an island with a rich historic past during much of the 17th and the 18th century.

It was not until the rise of the industrialization in the late 19th century that Dordrecht started to gain some development again. When the steam train networks started to develop, Dordrecht got connected in 1872 and got a bridge over the Oude Maas in 1885. The train station meant a new impulse for its urban development and some new neighborhoods developed outside the city walls. At the south side of the city new industrial harbors were added and thanks to its new infrastructures Dordrecht became the middle of a large industrial conglomerate of shipbuilding and metal industries and civil engineering around the Merwede and the Noord River. Nowadays the maritime industrial complex is the foremost economic asset of the Dordrecht and the surrounding towns that go together under the name of "Drechtsteden". The Nature of the economic activity has again enormously influenced Dordrecht. The rings of urban development and urban tissue around the historic core show the period of construction and economic activity by style of the urban architecture and types of buildings. The city reads as going from classic to modern as the distance to the town center increases. The residential neighborhoods are positioned towards the landscape and the industrial areas are situated along the river.

At the beginning of the 21st century the new architecture of Dordrecht excels in three directions. On significant focus is the surge of high quality apartments with views on the river s and the historic downtown on walk by distance. Living in an apartment or townhouse along the river or one of the canals is ever more popular. Around the Riedijks haven new apartment complexes have been built in fashionable contemporary brick architecture. Secondly a low-rise development trend that benefits of the local landscape qualities. A good example of more intimate development is project Plantij near the historic town center in an old water storage area. Here houses are standing with their legs in the water and the larger houses have a boating dock. Thirdly the transformation and revitalization of industrial heritage is increasingly important. A significant architecture is the refurbishment of an ancient energy factory and transforming it into a city theater. Another nice example is the villa Augustus complex where entrepreneurs from Rotterdam have opened a boutique hotel in a vegetable garden with hotel rooms in an old water tower. All these examples perfectly fit the rise of the leisure economy, which is an important engine behind urban transformations nowadays.

But maybe the most relevant contemporary development is the Dordrecht "Leer Park", a campus with schools for technical professions. It is a development that serves the maritime industrial complex by delivering students with high knowledge and hand-on skills. At the leer park regional educational institutions have been mixed with local companies and services in a park setting. The ultimate goal is to reinforce relations between companies and students to catalyze the knowledge economy and its monetary spinoff. This park is a relatively new development typology and will find many followers in the Netherlands. The urban design is conceived by the famous design firm West 8 and many Dutch architects have contributed one or more buildings. It will add a new layer to the rich texture of Dordrecht with its impressive historic past.

Almere
Newtown on new land

Almere is one of the few Dutch new towns. A town made on new flat polder land without palpable historic origin. For many years Almere was described as a remote anonymous commuter town without character and identity. Nowadays Almere has almost 200.000 inhabitants, a significant job market and offers a skyline, an attractive downtown, a central lake with cultural buildings and a large offer of varied and affordable housing. Step by step Almere is growing and maturing as a town and steadily it is becoming a brand of its own. Another leap forward, Floriade 2022 is currently under construction and serves to bring the town to international attention.

The story of Almere starts with one of the largest Dutch engineering projects in history called the Zuiderzee werken. In the 19th century plans were made to close the Zuiderzee to make the land behind the dikes safer and create better conditions for agriculture. The main supervisor and creator was engineer Cornelis Lely who designed and master planned the whole operation. He not only conceived the civil engineering side of the works but also had a distinct vision of the land for agriculture, leisure, production and new settlements. The "Zuiderzee werken", as the land and water planning project was called, had a scale hitherto unknown to the world. It was an amzing engineering project of creating a sweet water lake of 1100 km^2 where there had been a salty sea. Most important first step was to create a 32 km long dike to close the new lake called "IJselmeer" from the sea with the name "the Afsluitdijk". Secondly within the new sweet water lake five large land-winning projects called "polders" would be realized.

The "afsluitdijk" was finished in the 1930's and the new polders in the 1960's. Inside the IJsselmeer the largest manmade land is the "Flevopolder". Within this large new polder two new towns were envisioned with the names, Lelystad and Almere. Both cities are constructed two meters below sea level and protected by dikes of 6 meters high. Lelystad was named after Cornelis Lely the mastermind of the Zuiderzee werken. Almere was named after the ancient German name for the water meaning "big Lake".

Almere was conceived and developed as a suburban overflow town to cope with the rapid growth of population in the postwar era and the subsequent housing shortage in the region around Amsterdam. The construction of Almere started in 1976 with some small rows of houses in a big sandy plain. Many of its first pioneering inhabitants of Almere were people originally born from Amsterdam. A rapid expansion of Almere followed over the following decades and today Almere counts 200.000 inhabitants. Although the construction of Lelystad started earlier and Lelystad was planned as a larger and more important town, Almere quickly outgrew Lelystad due to its proximity to Amsterdam and the Randstad.

Originally Almere was planned as a poly-nuclear town with individual settlements in a green landscape. The idea was to bring the people in contact with nature and have a nice outdoor lifestyle. The first phases of developing Almere 70-80 % of the residences were constructed as social housing to make sure people from all social economic classes were able to move there. The housing was predominantly terraced housing with individual gardens and small scale apartments with a view on the open landscape. The planners thought that Almere could serve as a desirable alternative to the relatively bad quality of the housing in the Dutch inner cities in the 60's and 70'. The idea was that Almere as a newtown could compensate the lack of historic character and urban facilities of Amsterdam with an abundance of green space and wide views, affordability of the houses and the land, and a good car access. However, as is often the case, the planners proved to be shortsighted and as Almere grew and expanded as town, the ideology changed along with it. Nowadays the housing stock is hugely varied and the percentage of social housing is comparable to other Dutch cities. When Almere grew larger, the town's government realized that in order to serve the inhabitants the center had to become larger and denser with better and more facilities and services. So the poly-nuclear planning model was dismissed and the intermediate green space has been filled with developments to create a more centrally organized town.

One of the ways to win over the Dutch people, with their love for historic cities, was to apply special design themes for each new neighborhood. The idea was that strong or special design helps to create an identity. The theme approach served successfully as appetizer for outsiders to move over to Almere. Using Architecture and urban design to create an identity, Almere has become a collage of design trends. More than any other Dutch town, today Almere serves a showcase for contemporary planning methodologies. The theme design approach has been consistently applied for almost forty years now and has given special identity to the town, the town without a historic core. An early example of the special design-approach is the "Rainbow neighborhood" where different groups of buildings are colored according to the colors of the rainbow. Another example is the "Movie Neighborhood" where street names are called after actors and all the houses in clusters of 5 or 10 have a special and experimental architecture designed by famous architects. Many of the extensions and layers of development of Almere have served as testing grounds and exhibitions for new approaches in planning, development and architecture. For instance, to give the inhabitants more influence on the design or the architecture nowadays, Almere is a market leader in DIY urban development and house construction, offering its inhabitants opportunities for buying parcels and facilitating them to design and construct their own homes. Almere has one big advantage over all the other Dutch towns and cities. Given the abundance of cheap government-owned land and relatively easy construction in the spacious flat land, the housing price in Almere is still very affordable compared to that in the surrounding towns and regions even when ambitious architecture and planning is more expensive to design and build.

For a long time Almere was seen as a suburb for the lower middle classes that had been driven out of the cities by processes of gentrificatio and as an area for commuters that preferred to live in remote green areas with a good car access. But with the fast growing number of inhabitants and the steady increase of local businesses opening up, Almere gradually lost its aura of commuter town and made its decision to become a full service and mature town. A mature town comes with a healthy balance between inhabitants and jobs and a strong leisure and recreational potential. The city government realized that the most important step for Almere to become a full town was the creation of a downtown with a skyline, a retail area, a large hospital and a string of cultural facilities that need to be positioned along the central lake of Almere.

The city government also realized that for the branding of Almere as a Newtown assigning world famous architects to their important civic areas and buildings would be of a great advantage. After a strong competition the famous architect Rem Koolhaas was selected to design the master plan for the new downtown of Almere. The design is based on an ingenious idea to cross the public transportation lines in the downtown by introducing an artificial hill in the downtown. Koolhaas introduced the little hill with small slopes to cover the parking and bridging public bus lanes at the same time. The new topography gives a special feeling to the town center in the polder land where usually everything is flat. The retail areas are placed on top of the artificial hill. To make the plan likeable and to refer to the old dutch downtowns and city centers all the buildings and developments have been divided into smaller grains. The urban design gimmick here is ,however, that there is no application of classic architecture but only of contemporary design styles. This makes the new downtown in a visually lively and uncommonly cozy area when compared with other new developments in contemporary design styles.

Next to Koolhaas as a planner the town offers many architectural highlights from famous architects. One of the central blocks in the downtown with a mixed-use infill of residential, commercial and service facilities is designed by the leading French architect Christian de Potzamparc. Another architectural highlight is the theatre and community center positioned along the central lake designed by SANAA the leading Japanese practice. Having so many buildings by famous architects and experimental architecture gives a special feeling to the downtown areas of Almere. Although critics suggest the middle classes of Almere are devoid of cultural interest, one can see the special architecture applied as investing in the brand of the town and as a representation of its ambitions.

Since 2006 Almere has added a revolutionary tone to its palette of housing developments. In the Homerus neighborhood 3000 houses have been built in interaction with the future inhabitants. In varying degrees the inhabitants and homeowners have been involved in the programming, the design and the financing of the houses. Hitherto houses were always conceived by professional developers and the involvement of the inhabitants only started when they got the keys after signing the contract. This bottom up involvement of the home owners and users creates a new way of defining urban development by having a very detailed input in the formation of the area. Most direct form of this involvement is cutting out the middleman by selling the plots and parcels to individual owners and enables them to design and construct their own homes, in all budgetary categories, ranging from people with a small purse in terraced houses to the more well-off who can afford a freestanding villa on a large plot.

Given its abundance of green lands and plenty of space for extension in mid density, Almere is predicted to keep on growing to 350.000 inhabitants and becoming the fifth Dutch city. This, however, will still take one or two decades. Currently the newest neighborhood under construction is Almere Hout where all traditional ways of development and construction have been abandoned and the land serves as an open field for initiatives with a good plan and a financial backing. This kind of planning freedom is very unusual in the Netherlands and might become very popular if the new procedure shows good results and satisfied inhabitants.

In 2022 Almere will organize the Floriade, a floral and horticultural event with global aspirations. This event serves to promote the Dutch industry, the Dutch brands and to establish the organizing town as a place of interest. Almere has invited Winy Maas from the famous Dutch design studio MVRDV to provide an innovative sustainable master plan for the event. The master plan provides a spectacular grid folded over the highway reinforcing the existing infrastructure and city access. After the event the area can be easily transformed into a mixed-use district that also benefits from the initial investments for the Floriade. Again Newtown Almere is trying to bring itself to the next level.

ART TOWNS

Bergen
Birthplace of the arts

Bergen is one of Hollands most famous art towns of the 20th century producing a slew of famous artists under the name of the "Bergense School". Albeit a long and moving history where Bergen was a battlefield in three different wars, the most special features of Bergen are its specifics as a town located on the transition from sandy dunes to polderland with its subsequent variation of landscape beauty and vegetations. Nowadays Bergen is a famous art town and beach resort. Its many wonderfull villas in expressionist architecture are a testament to its artistic legacy.

Bergen is an exquisite little town in the province of Noord-Holland. According to historic sources the foundation of Bergen goes back to the year 800. It arose at the intersection of the sandy dune landscapes along the north sea and the north Holland Polder land with its green meadows. The meeting of different landscape types gives Bergen a special character where the transition of peaty soil to sand land makes for a rich diversity of vegetations, trees, and topography. The oldest and the wealthiest parts are located near the dunes since the land was higher and better protected from floods and storms and its vegetation offered better comfort for its inhabitants. Being in between the North Sea and the Zuiderzee there was always the danger of the water in the early periods. During the large and devastating floods of 1421 called the" Sint Elisabeths flood" that rearranged the map of the Netherlands, the neighboring town of Petten was totally washed away.

Due to its proximity to Amsterdam and Alkmaar, two of the classic towns of the Dutch republic in the 16th and 17th century, Bergen had it fair share of historic events, both good and bad. Towards the the Golden century (in the 17th century) where the Netherlands became one of the most successful countries in the world, Bergen benefitted by the dry-making of its surroundings called the "Bergermeer polder" paid for with profits from overseas trade. With the creation of the Bergermeer polder the future of Bergen and its water safety became more sustainable. Bergen suffered battles around Alkmaar during the 80 years war with the Spanish in 1574. Bergen was also a battlefield in 1799 when a large army of the English and the Russians tried to ostracize the French rulers from the Netherlands during the period of the "Bataafse republiek" and reinstate the royal family again known as the house of orange. Still a monument in Bergen commemorates the large numbers of Russian soldiers (more than 1700) that died in the battle of Egmond, far away from their motherland. But the role of Bergen as a battlefield wasn't over yet. The coastline of north Netherlands became of strategic importance in the Second World War when the German oppressors reinforced the coastline around Bergen with military facilities as part of a bigger project called the Atlantic wall. The Atlantic wall served as the eastern defense line for the German empire, and the area around Bergen came to serve as a military airport.

One of the major stories surrounding Bergen is about the Miracle of Bergen. In 1421 after the large flood disaster the people of Bergen found all the church relics of the neighboring town of Petten that had been swallowed by the sea. The church relics were reinstated and also the water inside the chandeliers was kept out of respect and superstition. When the priest checked the chandeliers after 6 months it turned out the seawater had turned into blood. In these ages this was seen as an intervention of the Holy

Ghost and Bergen became a place of worship. Many important people from the Netherlands and abroad came to pay Bergen a visit. To accommodate this huge influx of worshippers and to emphasize the new religious status of Bergen a large chapel was built next to the existing church and the complex was called the Pilgrimage church to honor all the pilgrims, (religious visitors) who came to visit Bergen. However, as a side effect of the raid on Alkmaar the church became a target of the Spanish aggressors in the 16 century. The church was partly devastated and only its skeleton remained. Nowadays this skeleton still stands in the heart of the little town and has the ominous name called "Ruins Church". The Ruins Church is still at the cultural heart of the little town and is central to events and manifestations.

However, Bergen didn't come to its greatest heights until the late 19th century. A few developments came together. As a product of the romantic movement of the late 19th century there was a strong re-appreciation of natural environments and small scale communities often idealized in romantic literature. This applied especially to the well-to-do classes of the larger cities. At the same time new technologies arrived like the steam trains and steam trams that were able to transport people outside of their town way faster than ever before. In 1901 Bergen got a direct train connection to Amsterdam and as a result the curious and culturally aware people of Amsterdam started to visit Bergen and especially artists settled down in Bergen. This can be considered the birth of Bergen as an art town or artist colony. The main attractions of Bergen were its leafy trees on its sandy grounds and the proximity of a dependence of the village on the seashore called Bergen aan Zee. The beach tourism had started in the late 19th century and early tourism facilities like small hotels had been settled.

The artist would locate themselves along the road in between Bergen and Bergen aan Zee called the Eeuwige-laan. A small district grew out of the artist colony with the name Bergen Binnen.

In the early years of the twentieth century the artist community organized themselves and gathered around artistic ideas and styles that later would gain fame nationally and internationally under the name of the Bergense School. The Bergense School with its focus on expressionistic and figurative art had many contributors grouped around a few central figures. The founder of the Bergense School was a French artist called Henri le Fauconnier, a Frenchman who brought with him arts theories from France. But the most celebrated artists of the Bergense School are the painter Charlie Toorop and the sculptor John Readecker. Readecker is the sculptor of the most famous sculpture of the Netherlands, the monument on the Dam in Amsterdam that commemorates the victims of the Second World War. The works of Toorop and Readecker have high values and can be visited in many of the Dutch top museums. The Museum for modern art in Alkmaar has a large collection of works of the Bergense School.

The artists who came to live in Bergen also brought their architects with them from Amsterdam. In this period just before the invention of modern architecture by the French architect "Le Corbusier" the Netherlands had developed an expressionistic and artistic architecture with bricks and tiles that was to be called the "Amsterdamse School". Subsequently the architects of the Amsterdamse School and their followers started to build houses and villas in Bergen in the areas near the old ruin Church and along the road to the beach. These villa areas have expanded over the years and nowadays are amongst the most beautiful and revered villa areas of the Netherlands.

The most famous villas are grouped in a cluster of 16 with the name "Meerwijk" that was commissioned in 1916 by the Amsterdam tile merchant Mr. Heystee. He approached the then famous Amsterdam Architect Staal to provide the master plan and the villa architecture. Staal was free to design what he thought was appropriate as long as he applied the tiles by the company of Mr. Heystee. Since it was a large effort, Staal himself designed five villas and brought a group of Amsterdam architects to design the other 11. Amongst these architects were the famous architects Kropholler and Piet Kramer. Kramer was one of the three leading architects of the Amsterdamse school. Together with de Klerk and van der Meij, Kramer designed the "Amsterdamse scheepvaarthuis" which is one of Amsterdam's most celebrated buildings. Later Kramer became the chief designer of Amsterdams construction department and designed many of the famous Amsterdam bridges of the 20th century. However, the villa park "Meerwijk" most significant and most famous villas are designed by the Architect Staal himself and the have the names like Villa "Ark" and Villa "Bark". In later years the works of Mr. Kramer in Bergen have unfortunately been lost. The group of villas was designed as an exposition of architecture and their composition of cane roofs, curved bricks and wooden accents was to become very popular and find a lot of following in the rest of the Netherlands.

Nowadays Bergen is a classic Dutch destination for beach holidays leisure and culture tourism. Its architecture and landscapes have been called national monuments and a visit to Bergen is sure to impress people with it local sensibilities. The dunes, the forests, the polders and the wonderful architectural heritage are testament to its status as an art town.

Laren
Arts and villas in the woods

The story of Laren is the story of a little Dutch town that was founded in the forest on one of the rare Dutch Wolds. Starting out as an isolated agricultural community Laren became an artist community in the late 19th century. The beauty of the natural environment and the typical farm-architecture proved very attractive. Famous dutch artists like Vincent van Gogh, and Piet Mondrian also spent time in Laren. The arrival of the steam-tram in 1883 facilitated the boom of the artists and made Laren into an early Dutch commuter town. Nowadays the cultural heritage, the lovely village center and the high level architecture of villas in the forest have transformed Laren into a habitat for TV stars and nouveau-riche.

In the middle of the Netherlands one can find a large landscape element from prehistoric times called the Utrecht Wolds. It is a remainder of the last glacial epoch where huge amounts of ice came from the arctic area's and reshaped the topography of the Dutch flat lands. This strip of low sandy hills stretches over a length of 50 kilometers in between the Ijsselmeer and the Rhine. The Utrecht Wolds, a push moraine count around 30 significant hill tops with heights between 20~50 meters above sea level. The Utrecht Wolds is easily identifiable by a vegetation and topography of forests and heat moorlands that is very different from the Dutch water landscapes that have a flat topography of peat land. Historically the area developed a pattern of settlements that strongly differs from the Dutch water towns like Leiden en Delft. The villages are mostly forest settlements with agricultural production and cattle farming at its economic core.

Laren is one of these settlements on the north edge of the Utrecht's Wolds near the ijselmeer. It's really intriguing that the current attractivity of laren with its hilly topography and beautiful forests has its roots in the geological processes of long time ago. Laren was for a very long time a marginal village in between a heat more and a forest with outspoken customs for breadwinning and local culture. It started out as a little agricultural community with no more than 50 farmers for a long time. A special feature of the region was that the land was used collectively by all the farmers. This lack of private farmland meant that the whole community was open and organized without gates and fences. This specific way of agriculture of common grounds also defined the layout of the village by an open space in between the farms. The name Laren is derived from Laar which means open space in the forest. In the morning and the evening the cattle would be collected on a large open space in the village. As time passed and the village grew, the open space in the village became surrounded with more farms and eventually became the center of the village. To provide shadow and some wood production the open space called "es" would be planted with big trees. In later stages of development this open space also became a marketplace a few times a week for selling locally produced goods. This village typology with a large open space with big trees in the middle became known in Holland by the name of "Esdorpen".

The village of Laren stayed an untouched rural community for many centuries with very little influence and exchange with the outside world and developed a distinct style in language, local dressing and farm architecture. The photos of this era show a beautiful, peaceful and almost utopian village where men and women are dressed in a special style. The farm building developed a vocabulary of its own. The farms usually have a length of more than 25 meters. The front part is a living area with spacious living rooms and the larger back part was meant for the cattle. Cattle and living parts were separated by a thick brick wall with a chimney that would be connected to the heating and the kitchen area. The farm roofs were made of canes that would isolate the home in the cold winter and keep the living quarters cool in summer. The farms all have an orientation that would minimize the forces of the strong autumn storm winds. The cattle would be led into the stall via doors from the long southside of the farm to minimize flows of cold air. Over time the long roofs of the farms with their wooden construction would sink a little due to the gravity. The shapes of the roofs became curved with the lowest points in the middle of the roof. This process of decay resulted in very picturesque architecture that would be appreciated by visitors from outside the village.

It was not until late 19th century that the then famous Dutch painter Jozef Israels found Laren and got excited by its untouched beauty. In his footsteps many fellow artist friends followed and little groups of them started to live in Laren. They bought themselves old farms and built new villas hence an artist community was establishing itself with diversity in caliber and standing of the residing artists. One of the most famous artists of that era also moving to Laren was Anton Mauve. Mauve was in his time an internationally well-known artist and attracted a whole new following to Laren. Next to his fame as an artist Mauve was also related to Vincent van Gogh and took the young van Gogh under his wings in his formative years introducing him to new skills and new directions in art. Also a well-to-do

American art collector called Singer fell in love with Laren and relocated there. The American family left strong traces in Laren in the form of a still well known and celebrated Singer museum for modern art. Another famous inhabitant of Laren was a young Piet Mondriaan who was on his way to become one of the most famous artists of the 20th century. Although the artists all gathered in Laren and got inspired by its natural beauty they never formed a well defined movement in history of art.

One of the reasons for Laren becoming a center of the arts around the late 19th century was new modern infrastructure. In 1883 a steam tram connection between Amsterdam and the little town of Hilversum was founded with Laren being one of its stops. A ride from Amsterdam to Laren would take less than 40 minutes, which implied a small revolution. Before the steam tram the basic means of transportation were still horse carriages and boats, neither of which was very fast. Highways and cars would not appear until 60 years later.

Laren, with its wonderful landscape not only established itself as an art town but in more general sense also as an commuter town and a recreational area for the more well-to-do people. The artistic interest and a relatively good wealth of Laren's inhabitants clearly show in many of the buildings and the architectural heritage. Laren counts a hundreds buildings qualified as national heritage and numerous buildings that qualify as local heritage. Next to the many farms this heritage also includes contemporary designs from the first half of the twentieth century. Laren had its own "society architect" by the name of "Wouter Hammdorff" who would design many studios, villas and also the Singer museum in an immediately recognizable romantic architecture that combines a modern approach to function with classic shapes and materials. Famous examples are his villas from the 20's that been the décor for many TV series, and have expression of convenient wealth in the style of Dallas and Desperate Housewives. Othere examples of his work are Laren's water tower which manages to combine elegance and functionality in the context of the forrest and the local topography.

Laren is the heart of a region that's called "Het Gooi". This region includes other towns like Blaricum and Hilversum which is the national broadcast town. Many famous TV personalities and influential people from the Amsterdam region choose to live in or around Laren. Many well-off people live in the Gooi because of the exclusive offering of villas in a wonderful landscape setting of tall trees and long green lanes. "Het Gooi" also offers many high end facilities like restaurants, galleries, boutiques and other services. Because of the specific population of "the Gooi" both Laren and Blaricum are famous as Holland's most wealthy towns. Another special facility is the "Rosa Spier Huis". It's a retirement home for artists and scientist that can get old in an artistic and productive environment while being taken care of. It's one of those special initiatives that fits the exclusive features of Laren as an art town.

Many towns with long historic roots have seen transformations in use while keeping its beautiful natural and cultural assets. So has Laren. From farming to art and from villas in the forest to hotspots of the nouveau riche, all is enabled by consistent quality of the landscape framework and love of cultural heritage.

PRODUCT TOWNS

Lisse
A Town that Guards the National Flower in Holland

People who have visited Holland in spring is likely to go or must go to Lisse, because it is the world most beautiful flower town, which guards and develops the history and glory of national flower tulips. There is also a well-known park Keukenhof, where visitors will have the fascinating visual experience.

Lisse is located in Western Holland with a long history, belonging to South Holland Province. The recorded history of Lisse has been more than 800 years or even longer. The experts speculate that the name of Lisse is likely to mean fence or railing, which may be the source of town square. This is related with the "Square Fortress" urban characteristics in Celtic Times, during which the Celtics like to build the square town in high places or along the waterways for commodity storage and trading as well as livestock protection. Now in some of the place names in France and England, Lisse has the meaning of hill and fort. Another statement is that Lisse is not fixed soil.

In middle ages, Lisse was still very small. There were only 50 households of residents around the year of 1500. Continuous wars led to poverty, and people at that time lived on peat exploration. Now Lisse has a population of about 22000 with the coverage area of 16 square kilometers, including 0.5 square kilometers of water area. The tulip based ball production and trade is the main industry of Lisse, which brings the development and prosperity of tourism and service industry relying on the world-known park Keukenhof. This makes Lisse a beautiful and vibrant town in Holland, continuously conveying the national flower spirit to the world.

Tulip - the gift from God

In 1593 when the first tulip was imported from Turkey, and successfully cultivated by the Dutch horticulturalist, this gift from God, the wonderful gorgeous flowers in the cup shape, makes the entire Holland crazy for its beauty.

Girls do not love the crown, sword and gold. Rather, they prey to the goddess to make the crown into flowers, sword into green leaves and gold into balls, which will make a tulip. At that time, tulip is not only the symbol of beautiful and romantic love, but also the keepsake for lovers. Poets and painters express their endless praise to this exotic flower. Dignitaries rush to take it. Thus, tulip also becomes a symbol of wealth and taste.

Unlike the romantic French, the Dutch are known for their adventurous spirit and business acumen. There are also some commercial speculators. While enjoying the romantic and beautiful tulips, they soon find its huge business potential. Those speculators even write romance stories about the tulip, and mock the British who cannot appreciate the tulip, as to stimulate people's desire. With Holland developed financial market and a large number of adventurous businessman, the tulip price reaches 10000 NLG after rounds of bull operation. Nowadays in a small town named Hoorn, there is a large house standing in the center, with three tulip stone sculptures in front. It is to commemorate the purchase of this house with three rare tulips in the summer of 1633. Later the house is converted into a Catholic Church. There is nothing special of this building, but it is the witness of tulip mania.

This "gift from God" not only blossoms its beauty, elegance and taste, but more importantly makes a lot of people respect and long for this town, which at the same time lays a foundation for financialization of contemporary economic products. Moreover, the tulip also enables commercial speculators to write a "Mad Dance", and performing the tragicomedy shocked the world.

Repent for the dishonesty

In the process of tulip bulb trade settlement between 1636 and 1637, the Dutch created the "futures option", introduced trading leverage and allowed overbuying. Nowadays the popular words in securities markets such as bearish, options and futures were created by the Dutch people at that time. On February 4, 1637, the buyer sold in a large amount, the market instantaneously went into the state of panic, and the price of tulip bulbs slumped. Subsequently, the tulip trade contract became worthless, and the contract holder began to ask for payment of the supplier. The previous booming scene suddenly turned into the hell with headwind and debts. The Dutch government cannot do anything except banning the tulip trade contract signing and speculation. Again a year later, the government regulated that the final tulip buyers can terminate the contract after paying 3.5% of the contract price. Since then, the Dutch tulip speculative market depressed. The historian Anna Gordon once wrote in the book Tulip Mania that 'the biggest damage of speculation is against the Dutch values. What's worse than cheating and forging?'. Let's take ourselves as liars. Tulip trading is not wrong in money, nor in business. It is wrong that people neglect the orderly society built on trust, which crushes the Dutch merchant class in economy, and accordingly business discredibility brings more fatal influence to the society,

destroying the whole economy in Holland and making it lose its economic supremacy. Although people from Lisse feel pity for the tulip suffering, they do not change their appreciation, respect and longing for tulips, nor give up and escape due to the tulip market depression. They choose to grow other bulb flowers such as HyacinthusorientalisL and import Hippeastrumrutilum from Mexico and lily from South Africa, extending the bulb flower production and development

People in Lisse insist on their emotion to tulips without hesitation, not only because of their love, but more like a kind of belief in God. They believe beautiful tulips no doubt belongs to the bliss garden, and are the gift from God. They need to not only protect the gift from the heaven, but also repent for the market discredibility which profanes the tulips.

People in Lisse are skilled with bulb flower plantation due to their persistence and relying on the suitable local soil and climate conditions. With much research and innovation, they develop more and more new varieties with different shapes and colors. People in the town guard the tulips through their faithful action, looking forward to its revitalization.

Keep Promise with Tulips for the Netherlands (God)

After the 19th century, the tulip production trade has recovered. People don't forget the lesson and place promise keeping over the trading business interests and value.
They attach great importance to the research and development of production and cultivation techniques, as to keep promise in terms of the quantity and quality. People in the town and other Dutch planters conduct the bulb flower production and technology research together. In order to guarantee the quality of the bulb cultivation, people take

advantage of the steam engine technology, and set up the water pump system in tulip planting areas, to control the underground water level and keep the stability of bulb field production quality. According to the characteristics of tulips, stimulate the bulb growth, accumulate the nutrient and reduce plant diseases, establish the high and low temperature plants, and research the bulb storage technology, to meet the storage quality and delivery ability of tulips and other bulb flowers, as well as to guarantee the tulip quality and the delivery time.

Sign the trade contract and keep promise backed by perfect mechanism. In the period of tulip mania, various nonprofessional speculators were involved in the bulb trade, which led to the bull campaign tragedy again. Later advocated by tulip growers in Lisse, the bulb production and trade is listed under the professional framework, and the KAVB (the Royal Dutch Bulb Growers Association) was founded in 1860. The bulb trade is based on the written contract, which is signed in accordance with the industrial bulb business. KAVB also sets the arbitration commission dealing with the relationship between sellers and buyers. Both sides have a chance to submit their disagreement and dispute to the arbitration commission. With the expansion of tulip bulb trade, the Dutch Wholesalers Association was established, professional bulb agent and bulb auction market mechanism were formed, which not only plays an important role in the business, but also charges the financial statements of trading activities, to ensure the trade and good faith. Bulb agent will determine the relatively stable market prices according to different varieties. Each year the KAVB and Dutch Wholesalers Association will evaluate the harvest results and remind the buyers in advance, so that the buyers can forecast the price variation in the coming year, which ensure the trade price and payment credit.

The expression of honesty of people in Lisse

Growers and researchers can control the bloom time through careful cultivation and storage. For example, suppose your birthday is October 17, and you expect the tulip blooming at that day. Then you tell the wish to tulip bulbs sellers and they will process the bulbs for you. After getting these bulbs, you grow based on the time and temperature as required. Till your birthday, you will be surprised that the tulips are blooming for you. This magic cannot happen in other flowers. This fantastic skill, professionally named forcing culture, is regarded as the promise by God. Gardening scholars from all over the world are wondering about this. Once I accompanied the domestic flower experts to visit the bulb manufacturers in Lisse. While talking about the forcing culture, the hosts said humorously that it was not a technology. Rather, it was the dialogue between bulb growers and tulips, and the response of people in Lisse to keep faith.

Let the National Flower Blossom More Brilliant

In the ten years after the World War I, bulb trade gets unprecedented booming. The key driving force of the success is the international scope of professional publicity. Tulips can be seen everywhere in the world famous parks, exhibitions and various gardening associations. Since then, the foundation of tulips as the Dutch symbol has been laid. Today, tulip almost becomes the synonym for Holland. Tulip themed cultural exchange activities are held throughout the world. There is no Dutch themed activity without tulips. Tulips as the symbol of Holland firmly root in the heart of the people all over the world. It is evident of the cultural functions brought by bulbs represented by tulips. Lisse town, which starts from tulips introduction and guides for nearly 400 years, is the most faithful representative for the Dutch national flower.

Whether by bus or by bike to Lisse, you will see a small wheel in the center of N208 main road and branch. On the central flower bed, there is a tulip sculpture with a clever and simple design. The branch at the intersection is the gateway to Lisse.

The planning layout of the town is closely related with the function of production and trade of bulb flowers, with reasonable arrangement. The town center is composed of two main streets in the cross shape. Streets outside the center are scattered with the houses and bulb production and trade companies. There is a canal in the middle with its outer areas connecting the bulb production company's huge processing and storage warehouses. The periphery of the whole town is surrounded by the colorful flower field like a carpet. In the core area of the flower field, there is a tulip park that fascinating the world.

In Lisse, you will see varied bulb flower images represented by tulips. Each street in the town is named after the bulb flowers. Walking around the town center, regardless of the street greenbelt, roadside greening, and even residential garden along the street, you can see colorful flowers. All kinds of goods from oil paintings, carpets, furniture and wallpaper, to small vases, ceramic tile, silverware and wooden shoes, and to bags, ties, umbrellas, stamps and food, can be seen in almost all small stores. From these experiences, you can feel the tulips everywhere: the street furniture, building facades, grounds and even the building maintenance board. In the spring, whether riding or walking, you can put yourself in the colorful world full of tulips in the streets and feel the colorful dream world.

After the Second World War, flower tourism has gradually flourished. Since 1947, Lisse has had the parade on the last Saturday of April every year to express its thanks for the luck and wealth brought by Flora to the people in this town. People there also planned to build a bliss garden, as to show the natural most charming miracle tulips to the world. Advocated by the mayor in 1949, and based on the Keukenhof garden designed by the German landscape gardener in 1830, the Keukenhof garden was transformed into an open show base presenting the outdoor bulb flowers.

Since then, a worldwide gorgeous beautiful park Keukenhof appears in front of the international people.

Flower and bird viewing: The tulip varieties, quantities, quality and decoration ways are the top in the world. There are more than 6 million flowers in the garden with many rare varieties, and over 2500 trees of 87 species. There are also more than 30 species of birds and animals. Swans are swimming freely in the pool, peacocks are wandering on the lawn and squirrels are jumping happily among the trees.

Design layout: British style is given the priority. There are 7 different themed parks with different characteristics. For example, the Inspiration Garden is composed of varying gardens with their own characteristics and themes from classic to modern style, from which people can learn. Art Park contains hundreds of sculptures and works of art by the Dutch famous artists, and the natural elegant environment sets their beauties. There is a daily flower show Orange Nassau Pavilion and a weekly flower show. Willem Alexander Pavilion is the world biggest lily show each year. Themed flower parade has different themes each year with millions of flowers through Keukenhof Avenue. Visitors can learn the gardening knowledge through videos, pictures and interviews. In Keukenhof Garde, there is also a photography studio where people can learn practical photography skills. Children can learn flower knowledge and play on various amusement facilities in Children's park. There are also touring routes specially designed for children's adventures, garden maze and pasture, etc.

Characteristic activity: Since 2006, Keukenhof Garden has developed a new tradition that the garden will have a theme each year such as Chinese Olympic Games in 2008 and British style garden in 2013. This style attracts passengers and keeps a large number of visitors each year. On the first Saturday in May there will be a flowers mosaic exhibition and competition, during which hundreds of works are presented and each work is composed of thousands of flower petals.

Tourism node: Visitors can rent a bike and cycle in the flower field, visit the black tulip museum and stop to take pictures. They can also boat among the flower field to view the scenery from another aspect.

Today, the tulip garden in Lisse has totally received more than 60 million visitors from all over the world. The gross domestic production of tulip bulbs has reached 540 million Euros and the total export has reached 500 million Euros, to over 100 countries and regions of the world. Lisse, a small town featured as tulips in the Netherlands, is using the gorgeous brilliant Dutch national flower to decorate the living environment, and to maintain the Dutch integrity spirit with their persistence.

Alkmaar
Beginning of the Victory of The Dutch War of Independence

The statement among the Dutch people "Victory Starts From Alkmaar" is originated in the 80 years' war between Holland and Spain, telling a heroic story happened in the town of Alkmaar.

In 1572, the Dutch launched the civil uprising against high taxes and high centralization of the king of Spain as well as the Spanish brutal crackdown against Protestant. Alkmaar was no exception. In August 1573, to suppress the Holland uprising and revolt, thousands of Spanish soldiers led by Fernando Alvarez de Toledo surrounded Alkmaar after capturing several cities successively. The brave people of Alkmaar threw hot tar at the enemy and burning branches from the wall. That was one of the most glorious moments in the history of Alkmaar. The Spanish army hoped to trap the rebels in the city through cutting off the food supply. A brave carpenter successfully sent the aid letter out of the city. Fianlly the polder water level around Alkmaar was increased, and Spanish soldiers escaped. This is the origin of "victory starts from Alkmaar". It is the first time that the rebels led by Holland Prince William defeated the Spanish army, and Alkmaar became the first city in Holland breaking away from Spain. October 8th, the day that siege ended, became the celebration day of Alkmaar. Then the Dutch continued to resist Spain, and ultimately won the national independence after 80 years of war.

Alkmaar is a city with splendid history and rich culture. There are rivers crossing the town with a canal through the town center. The medieval buildings are reflected in sunlight flash in the river. The breeze makes the river sparkling with small waves. The ancient town Alkmaar presents a glorious and ancient charm. The golden age of Alkmaar starts from the St.Laurenskerk cathedral, located in the center of Alkmaar. Designed by the architect AnthoniusKelderman from Mechelen and built in 1440~1512, the cathedral is in Brabant Gothic style, with 26 columns to support the vaults and ceilings in the building. The rigorous structure together with the gorgeous appearance and elegant carving makes the church magnificent and solemn. This church is not only the representative and example of classical architectures in Alkmaar, but also the symbol of prosperity in the golden age of Alkmaar.

Developing from a fishing village in the 10th century, Alkmaar was granted as a city in 1254. Trade and waterway transport hub for centuries brought prosperity to the people in Alkmaar. After 1630, with the drying of several large lakes in North Holland, Alkmaar lost its connection of the South Sea and the maritime trade gradually reduced. A large area of polder was increased around the town. The humid land in polder areas and the rainy and foggy climate in the town favored the succulent herbage growth. Alkmaar previously relying on the trade transferred to the husbandry industry. The trade experience made the people of Alkmaar aware that the market was the key to industrial development. Carrying out a comprehensive production in the husbandry industry, Alkmaar tried to get the dairy products trading rights from the country and finally obtained the authoritative measurement tool—cheese weighing in 1365. Later the cheese market was established in the center of Alkmaar. People again developed the cheese market into the largest one in Holland with their business wisdom and effort, to maintain the leading position of cheese trading in Alkmaar, which won the reputation of Cheese Town and also favored the sustained prosperity and development of Alkmaar.

People in Alkmaar advocates freedom, and has the valiant and struggling character and culture. Alkmaar is the first town in Holland to advocate the belief of freedom and to achieve the coexistence of multi-religions. Before 1520, people in Alkmaar were catholic. Shortly after the St Laurence Church was built, more and more people hoped to reform the Catholic and established the Protestant. Under the leadership of Prince William, people who advocated the freedom of religion belief had the disputes with Catholics. The Protestants occupied Alkmaar cathedral and abbey, and established new rules and rituals. After that, the belief system of "one city, one religion" in Holland was broken. The multi-religions coexistence system was initiated by people in Alkmaar.

The story that people in Alkmaar defeated Britain and Russia expeditionary forces in the battle of Castricum is written in the Arc de Triomphe, reminding people of the brave people from this town.

Walking on the street in Alkmaar, people can enjoy not only the cheese aroma, but also the environment of different museums and exhibitions, which makes the whole town full of culture atmosphere. There are more than 400 state cultural relics. The elegant and magnificent medieval buildings in the town and crisscross ancient canal complement each other, making the charming scenery. Nowadays the most attracttive thing to the tourists is the traditional cheese fair show from 10 to12 am every Friday from April to September, following the three-to-four-hundred-year cheese trading history. There is no machinery, no cars, no computers, no fax machine and mobile phones, and even no people using the calculator. The cheese fair show can take the tourists to the time of 400 years ago.

Alkmaar cheese market is in waagplein square. Workers start to place the cheese at 7 o'clock in the morning. Then there will be about 30000kg (about 2400 pieces) cheese in the square. If it rains or the temperature exceeds 28 degrees, put a layer of special canvas below the cheese. The cheese porter must arrive before 9:30. People who are late or violate the rules will be publicized and fined. The fine income is partly to support a school in Alkmaar, a city of Surinam in the same name, and the remainder is used as the income of the cheese porter. Each porter is dressed in white and wearing the hats in four colors red, blue, green and yellow, which with the cheese wooden color can distinguish the four leading storage companies. The only one with the orange hat is the market agent of these four storage companies, known as "the father of cheese". Before marketing "the father of cheese" will tell porters the daily trading volume, important guests, and the carrying range of each company that day.

The bell rings at 10 in the morning, marking the official start of the cheese market. In the opening ceremony, film stars, famous athletes, foreign diplomats and other celebrities will be invited as the host, which greatly attracts visitors.

Judge the cheese flavor, fat content and hardness by seeing, smelling, knocking, cutting, drilling, twisting, tasting and other similar ways. Observe the inside holes and degree of uniformity through cutting. The inside holes are caused by beneficial lactic acid bacteria during the fermentation, which should be uniform in very perfect cheese. After inspectors' identifying the cheese quality, porters will move the cheese from the square to the weighting room, as to check the cheese quality. Buyers and sellers clap hands to bargain and determine the transaction, not needing to sign a contract. Then the cheese is weighed and marked. After the transaction, porters become the protagonist of the game, with one in the front and one in the back. They shoulder the basidium placing 8 round cheese, swinging hands and shuttling between the weighting room and the square. The exaggerated funny action becomes a kind of scenery that tourists favor most.

Another highlight is is Holland beauties in traditional costumes of pretty white cap, red scarf and blue apron, who wear wooden shoes, carry baskets filled with bags of cheese and sell to tourists. In addition to viewing vivid cheese trading, people will also taste this local specialty. After the trade, the cheese will be transported to the whole country and the world, spreading the cheese aroma to people faraway.

The cheese market will attract 300 thousand tourists every year.

Nowadays, while wandering in the center of Alkmaar and listening to those buildings telling the town history, people will be moved that victory still begins from Alkmaar.

Delft

Delft, located between Hague and Rotterdam, is the most elegant and featured town among all Dutch towns. Like most Dutch towns, Delft takes the church as the marker and center, and extends outward. The square in the center of an ancient city is composed of old and new churches as well as the city hall. The square primitive environment with the solemn and majestic feeling implies the relationship between the ancient city and royal family.

In the Spain Dynasty rule era, the first king of the Dutch Republic Willem van Oranje led the Dutch people to resist the Spanish for decades, and was stabbed to death in the royal garden of the ancient city. Later the new church in the centre of Delft was selected as the resting place for Willem, and then other royal members were also buried here. The ancient town originated from a man-made waterway named Oude Delft. The verb of deft delven means digging. Matching its meaning, Delft is an ancient town with man-made waterways and dikes to hold back the North Sea, drain the coastal marsh and form the ring of the city. This fully reflects the struggling and unrelenting spirit and character of people living there to fight against foreign enemies and nature.

Delft is a town with culture and stories. One of the most famous Dutch painters Johannes Vermeer lived and painted here throughout his entire life. The sunset shines warmly at dusk, with red clouds blotting out the sky at random. Besides the sparkling water, Delft is in the most beautiful time of the day. The picture is like the landscape described in Vermeer's paint of Delft Scenery: meandering waterways, white suspension bridges and green oaks, presenting incomparably beauty under the sun. The city shops and galleries, together with regularly held cultural activities such as the music festival and theater festival, tell the splendid history, art and culture of the town.

In the past, the Dutch usually procured the blue and white porcelain from China by the East India Company. However, it interrupted due to the frequent civil war in mainland China in the late 17th century. The shrewd Dutch began to study and imitate Chinese pottery and porcelain through making clear of the proportion of cobalt and manganese. Being not satisfying with what was obtained, the Dutch got the essence from Japanese glaze technique, integrated with Chinese ceramic art and formed the world-famous national industry and fine art products—Delft Blue porcelain (Delft Blue).

While exploring its reference significance to the feature town in China, we focus more on how the foreign industry carries forward and becomes enduring for centuries, rather than the generic process of blue porcelain. Obviously the local culture injects the foreign industry with inner soul, and makes it root and grow in the local environment.

The foundation tone of the blue porcelain culture is blue. Growing up along the North Sea, the Dutch fought with the sea and utilized the sea for a long time, during which the color of blue has been deeply engraved on their nation spirit. It is regarded that the Netherlands is a country passionate with blue, which was covered by the beautiful and emotional orange tulips.

Blue porcelain products are from daily life in the Netherlands. Typical things such as windmills, wooden shoes and tulips are the main characteristis in those products. Rich and colorful images on the porcelain are the microcosm of Dutch 700-year cultural life and art creation. It is not strange that a product that can reflect the national soul and engraved with life features becomes the world-renowned national products.

The blue porcelain also reflects the innovation spirit. To solve the light difficulties in Rembrandt's painting Night Tour, it takes more than one year to adjust the pigment allocation and make 480 pieces of porcelain with different colors. The customized production of blue porcelain, especially the royal order, or the gift as the "national gift" to other heads of state, greatly enhances its brand image and the grade. Tourists can also make their own blue porcelain through imagination, and experience the fun. All these can be learned in the industry development of Chinese feature towns.

Delft is filled with the atmosphere of primitive simplicity and elegance, as well as vitality.

Walking around the town in the morning mist, you will see many wild swans, seagulls, ducks and geese swimming in ancient city rivers. Sitting beside the ancient canal, and watching the white swans and ducks asking for food from people. What a harmonious picture of humans and nature! The vibrant environment there is also related to the integration between the Delft University of Technology and the ancient city. Ranking No. 65 in the world and No. 5 in architecture, planning and environment disciplines, the Delft University of Technology has an open campus, which separates from the buildings. The university is a part of the city space. The urban streets, commercial and other public facilities provide services for both the university and urban residents. The college students and citizens live in the city together, where the knowledge and experience, history and modern things collide, to merge into a new culture and continue in this ancient town.

Streets and roads in this ancient town are pleasant with the width of about 5~8m. Buildings in typical medieval and Renaissance style on both sides stand along the ancient canal. The main color is brown, mixed with white and orange. The architecture details vary and are finely crafted, with museums, memorial halls, antiques and porcelain stores among those buildings. It is warm and pleasant with lasting appeal. The road in the middle is for the pedestrian and cycling, and cars can be seen only sometimes, which gives people the feeling to shuttle between the ancient and modern times. This is the original Woonerf Principle created by Dutch people, creating the safe, integrated, interesting and harmonious traffic environment between people and streets.

The current development of Delft is taking EWI electronic building as the highest point and Aula, the central library and the new railway station as the new landmark. The Aula building designed by the Dutch architect Jacob Bakema is known as the UFO due to its flying saucer shaped large cantilever, which is also regarded as the representative of straightforward concrete building. The central library and the new railway station are designed by Dutch Mecanoo Design Company. The library is famous for its large ecological slope roof, which embodies the symbiosis between humans and nature. The newly designed railway station is over the concrete elevated bridge, and the outer layer glass reflects the sky, which becomes a perfect example of old city transformation combining with the new city. The new landmark buildings are good reflection of the town from the ancient to the modern times. They are developing naturally due to the need, carrying the inherent genes of the town and instilled with the straightforward, challenging and innovative spirit of residents. They are like the children cultivated by the old town and now have become a part of the city. This development mode based on requirements is worth considering and referring to in the Chinese feature town development.

Enschede
Resilient after disasters

How can city survive disaster or may be even two disasters? Enschede is a Dutch town in the east of the Netherlands that resiliently managed to deal with disaster. First by changing its entire economy after its abundant factories were outsourced to low wage countries and secondly when fireworks storage exploded and blew a hole of 63 hectares just north of the Town center in a lively district. Enschede managed both challenges and set examples for diversifying its economy and rebuilding for the people by the people and with the people after disaster.

Rural settlement becomes a town
Enschede is not a typical Dutch town since it doesn't have a river or defense works. It came up late 19th century as a center of industrialization and textile and beer production. As town it grew on the crossing point of regional roads intersected with an early train line connecting Amsterdam and Rotterdam with the European hinterland. The blossoming period of the local industry reached its peak somewhere in the 1930's and many parts of the downtown are formed in this period. Next to the round historic center the beautiful 1930 Singel circles the city and provides a strong frame for understanding and experiencing the city.

Center of industrialization
The industrial development made Enschede the early center of Dutch industrialization. Famous are Enschede historic images of large factory complexes, smoking pipes and strong productivity intertwined with a rural landscape structures. Also Enschede is lodging some of the finest examples of workers housing in the form of garden villages, still impressive and popular with names as Patmos and de Laares. The textile barons of Enschede had a large share in commissioning workers residences and beautiful city parks.

Factory Foreclosures

However starting in the 1960's many of the factories closed and left the 30.000 employees out of work over a period of two decades leading to sever social crises and difficult circumstances. Many of the factories were removed, converted or simply abolished. However at the same time in the 1960 sixties Enschede was awarded a university from the national government that went on to become the technological powerhouse nowadays called Technische universiteit twente.

Economic diversification

The university with its large of spin off of commercial application in technologies has provided Enschede with a new economic perspective. The university and it's allied companies and institutions are Enschedes largest employer today and attract many workers from the northern region of the Netherlands. After the new football stadium was constructed it looked like Enschede found its way up again and was ready for the 21st century. However…

Fireworks explosion Disaster

On the 13th may of 2000 disaster struck again. A fireworks factory and storage facility blew up. The explosion of the factory, located in an old industrial mixed use neighborhood with many residential units, left many dead and 1000's wounded and struck a gaping hole of 63 hectares just north of the downtown. It was one of the largest disasters taken place in the postwar Netherlands. After visiting the site of death and destruction the then prime minister of the Netherlands Wim Kok made the promise that nobody would be left worse of and funds and special powers were allocated to reconstruct and improve the disaster stricken area.

Leader for the reconstruction

One of the main questions after the smoke had cleared who would be assigned to coordinate and design the reconstruction of the town. By the end of 2000 the revered Dutch architect Pi de Bruijn a native of Enschede and famous as architect of the national parliament and urban supervisor of large projects in the capital Amsterdam was assigned to led the reconstruction. De Bruijn, was chosen based on this brokering skills between government and commerce and his capacity to involve the people of the city in the planning process.

Building with the people for the people

Since many of the people who lost their homes still owned the land on which there houses once stood a unique process was conceived where large developers were left out and the people of the city could built by themselves on their own plots. Since Netherlands has a strong tradition in collectively developing their cities a unique approach was born. Almost 1800 house would be built by the people themselves and the involvement of the people was organized on three levels in the process of planning and design.

Interactive development

Before any design was made all the inhabitants and stakeholders were interviewed and asked for their preferences in the urban design and the living conditions. These results were used as an input for the redesign concept. Secondly the urban design concept was evaluated by the people of the city and openly discussed and improved and finalized. Thirdly the people were able to build their own house on their own plot. The strategy was approved with popular votes mainly because there was no radical new design but all the remaining elements where carefully crafted into the lay out of new plan so the memory of the place would be enhanced by the contrast of the old and the new constructions of the town.

Building for all

In the urban plan conditions were created for the social and economical mix of the local people. Some of new area consisted of small plot where people could construct smaller houses wall to wall in a terraced typology. In other areas with larger parcels houses could be larger and freestanding. This mixed allowed for a large variety of peoples and companies to construct their own house or facility. Along the way it become clear that the ability to construct your own home is a very strong impulse for community development and create added value in the beauty of the built surroundings.

Creating a rich texture

As a complement for the residential and commercial facilities a layer of cultural and educational institutions was added to energize a flow of visitors to the area. The public space was designed with inclusion of the trees that survived the disaster. The reconstruction process almost took 10 years to be completed and in 2010 the reconstruction was declared finished.

Setting a new Dutch standard

Nowadays the roombeek neighborhood enjoys a huge popularity and numerous busses of visitors visiting the area year after year. The planning strategy of locally involving the people and have them building on their own plots without involvement of institutional developers has attracted a huge following in the rest of the Netherlands. Nowadays almost all new developments carry some of the planning methodology as developed in Enschede. One of its foremost followers is the newtown of Almere where the mayority of residential units is constructed without developers by happy people building their own home.

The disaster laid bare a strong resilience. A wonderfull neighbourhood was constructed and set an example for the rest of the country. Enschede can look ahead again.

SCIENCE TOWNS

Leiden
City of discoveries

Leiden is one of the famous Dutch towns from the Dutch republic in the seventeenth century on par with Delft. Its streets, squares and canals are organically intertwined with its long and rich academic history of more than 400 years. The city is sprinkled with institutions and facilities for education and culture. Nowadays an impressive center for biomedical sciences has sprung up in addition to its university. For its residents and for families with growing children Leiden is a comfortable and relatively quiet town with excellent facilities in the Dutch ring of cities called the Randstad.

Leiden, like many other classic Dutch towns is founded on the junction of two rivers, the "Oude Rijn" and the "Nieuwe Rijn". At the junction small settlements arose for trade, production and administration that were to form Leiden. From 15th century these little settlements merged and transformed in a walled city with Holland's most important waterways flowing within its walls. The water connected Leiden with the North Sea at nearby Katwijk, with Utrecht in the east, and with Rotterdam in the south. At the crossing point of trade routes, trade flourished and Leiden became an important industry center around the production of linen. Leiden also played a pivotal role in the 80 year war with the Spanish kings from 1568~1648. Resisting the Spanish oppressors Leiden was set free in the early phases of the war on October the 3rd of 1575. This was a small but important step in the formation of the Netherlands under the supervision of Prins Willem van Oranje (or as he is called, Willem de Zwijger), in later times regarded as the founding father of the Netherlands. The third of October is still celebrated every year by the people of Leiden as the "Leids Ontzet".

Leiden lacks squares and markets that can be found in other classic European towns. Its memorable spaces are still the urban waterways of Oude Rijn en Nieuwe Rijn with their joyous curves flowing along the "Castle of Leiden" at the heart of Leiden's historic center. Circumscribing the junction of the rivers are the Oude Vest on the North and the Steenschuur on the South. The Steenschuur flows along the most famous street of Leiden called the Rapenburg which is curved around the main cathedral called the Pieterskerk. The Pieterkerk is amongst the oldest and largest structures in town dating back to 1390, and constructed in a late gothic architecture. Opposite of the Pieterskerk on the Rapenburg one will find the most famous buildings of Leiden called the Academy building (academiegebouw). The Academy building has been at the heart of the university for centuries. Its first foundations were laid in 1516 and the building started out as chapel for a cloister of Dominicans, or the white nuns. The building is organized in eight bays and has high glass windows conceived in a pragmatic Dutch architecture. The most famous space in the building is called the "sweat room" where all the recently graduated master and PhD Students are invited to write their signatures on the wall. Highlights are the autographs of the previous queen Beatrix and the current King Willem Alexander. Traditionally members of the royal family will study at Leiden University.

Leiden University was awarded to Leiden in 1575 by the Prince of Orange for its military persistence towards the Spanish aggressors. It was not only a gift but also a necessity, since the Dutch republic would benefit from well educated citizens and good education was until then to be found outside the Netherlands. Early universities were founded in the 13th and fourteenth century in Ghent en Bruges, the region we now call Belgium. Founding the Leiden University coincided with the rise of scholarly science as a product of the renaissance that spread in Europe in the 16th century. It was not long before Leiden was considered a leading academic institution in Europe. Famous scholars of the early days are, for instance, Hugo Grotius, the inventor of international law, and Gerardus Vossius, one of the first scholarly theologians and classicists. Another towering figure from Leiden University was Hermann Boerhave who is universally seen as the founder of modern medicine and has his own museum in Leiden. Although most people have no clue anymore who these scholars were, many buildings, streets, and organizations are named after these famous scholars from Leiden and they live on by the use of their name.

The university grew with ups and downs but was steadily on the rise in size and reputation. In the late 19th century and the early 20th century the university had a prodigious output in the field of physics and chemistry with famous physicists and chemists such as van der Waals, Lorentz, and astrophysicists Zernike. At the world's first university low-temperature laboratory, professor Heike Kamerlingh Onnes achieved temperatures of only one degree above absolute zero of −273 degrees Celsius. In 1908 he was also the first to succeed in liquifying helium and can be credited with the discovery of the superconductivity in metals. The reputation of the university was such that giants of science like Einstein, Ehrenfest and Freud all spent significant periods living and working in Leiden in the first half of the 20th century. Nowadays

Leiden University is considered one of the top universities of mainland Europe. The university has 25,000 students and a staff of 3000. Amongst its peers are 16 noble prize laureates.

During the course of its development Leiden University spread its buildings and its institutions around the city mixing with the ordinary life of the citizens of Leiden. The planning of the town and the university go hand in hand and the presence of many students gives flair to the town. Many of the impressive academic institutions have been integrated in successive periods of urban expansion. The early buildings are all integrated in the historic core. Just outside the historic core in the 19th century development zones many large classic buildings can be found like the Huizinga building, the university library and for example the university museum of ethnology. The defense zone around the historic town center of Leiden is still intact and provides a glorious view on the historic town center. The former defense works have been readjusted into a series of parks surrounding the historic downtown and providing the university buildings and institutions with magnanimous green walking and gathering spaces. Keeping the defense zone and not constructing an inner ring road like what happened in many other cities results in a difficult car access to the historic downtown and its university facilities. Since the early 20th century Leiden has expanded considerably beyond its historic borders and 70% of Leyden's buildings are constructed after 1945. In the third wave of growth, new university areas have been created of which the current bioscience centre north of the train-station is the most representative one.

This bioscience park holds a potent mix of commercial research and scientific institutions that fertilize and accelerate each other. The bioscience center is pivotal to the economic offspring of the Leiden University of life science innovation and a large publishing industry. The bioscience park is a technological powerhouse with innovations that connect to all major economic areas in the world. In this park the most impressive works of contemporary architecture by leading Dutch architects can be found. Ben van Berkel of UN studio designed the Mirai building on the bioscience campus that serves as headquarters to pharmaceutical multinationals. Erick van Egeraat designed the new buildings for the Heike Kamerling institute for physics and astronomy. Leiden harmoniously balances classic heritage with contemporary architectural development. Next to the construction of the institutes and academic buildings Leiden had a little boom in museum buildings, which fit the scientific character of the town and which provide leisurely educational entertainment for the inhabitants of the Netherlands. The largest new museum for Leiden will be the reconstruction of the Large Naturalis museum by the architects of Neutelings Riedijk, famous for their celebrated museum designs all over Europe. These museums amplify the city brand of Leiden defined by the slogan "City of Discoveries".

Reconstructing the History of the University one shouldn't overlook the fact that Leiden also features many convenient neighborhoods for its more than 100,000 inhabitants, of whom many are not related to the university and its ecosystem. Living in Leiden is also appealing by its nice historic feel, its convenient neighborhoods and its rich facilities and services that come with being a university town. Living in Leiden is foremost attractive by its proximity to the A4 highway and the national rail service that serves as economic arteries of the Randstad Metropolis. Located in Leiden one can reach the major cities of Rotterdam, Den Haag and Amsterdam in less than 30 minutes. Schiphol, the national airport is less than 15 minutes away by train and by car. A neighborhood that exemplifies these attractions is the "burgemeesters and professoren wijk" (the Mayors and Professors district) that is designed and built in the 1930's adjacent to the historic center. Its spacious layout and chique architecture in combination with proximity to the center and the facilities makes it a very popular area for family living. The unique selling point of Leiden is its high convenience within a small scale historic living environment in a relatively short distance of all major economic hubs of the Netherlands. One could state that Leiden is a town attractive to people with a regional or national reach that cherish at the same time comfortable and familiar surroundings. It is a good alternative to larger cities, especially for people with growing kids that need a safe and comfortable living place with a cost that is significantly lower than that in cities like Amsterdam.

The most interesting contemporary construction project appealing to the vanguard of inhabitants is a neighborhood called Nieuw Leiden, a high-density low-rise neighborhood placed within the historic boundaries of the city. The urban design is based on a grid within which the citizens of Leiden can acquire a plot and build their own houses. In this experimental neighborhood building by project developers is exchanged by individuals building their own habitat. The construction of individual houses on individual plots is guided by the rules of the master plan for the area. The guidelines and the master plan are conceived by the famous Dutch design firm MVRDV. The guidelines provide an harmonious development process where each individual building contributes to the beauty and the wellbeing of the whole. Most of the houses are terraced houses with small patios so a relatively high density can be achieved. The architectural styles are all formulated under the umbrella of progressive architecture with natural materials and refined detailing. Also the aspect of sustainability is addressed by circular approach to flows of waste and energy. This development in Leiden is representative for a large wave of similar developments in the Netherlands where citizens can realize their own house in an urban setting. An increasing number of people is enthusiast about this approach that seems a credible alternative to buying houses from developers with their preconceived floor plans and uninspiring facades. The recent economic crisis accelerated this type of citizen based organic town development since large scale capital intensive projects were put on hold during the banking crisis. Most self realized houses in Nieuw Leiden define new typologies for family living and will inspire people from other cities to follow in their footsteps.

The historic downtown, the beautiful canals, the new museums, the self-realization neighborhoods, the university and its ecosystem, all add to creating an excellent centrally located service town full of characteristics. Leiden, a city of discoveries, makes people discover they don't need to live in a large city .

Wageningen

Wageningen is located in Gelderland province, the middle of the Netherlands, with an area of 32 square kilometers and population of 37000. It is a European and even world-famous "life science city".

Wageningen is a famous historic city in the Netherlands, with many medieval early churches and forest ruins. Waglingenburg and Strasbourg are the main castles in the city. In 1263, Wageningen acquired the city rights. In 1918, the first educational institution Lamborghao School was founded there, which is the predecessor of the University of Wageningen. Since then, the town has started the journey from an ancient small town to a modern agricultural science and technology park as well as the Food Valley Center.

Wageningen is known for its historical events. On May 5, 1945, the General Foulkes on behalf of the Allied Army negotiated with the German General Blaskowitz in Wageningen World Hotel, and signed the German surrender book. This means the end of the Second World War for the Netherlands. Therefore, Wageningen is also known as the "Liberation City". That is why May 5 is known as the National Day of the Netherlands. Other European countries select May 8 as the National Day, the day when Germany officially announced the surrender.

It is well known that the agriculture in Holland is at the world advanced level and 56% of the total land area in Holland is used for agriculture. The agriculture in any country of the world cannot be compared with that in Holland in terms of the efficiency, production and innovation. If farmers in all the other countries are as productive as Dutch farmers, then 1/4 of the world agricultural land is only needed to solve the food problem. The cow in Holland has high milk yield. However, the resulting greenhouse gas emission is lower than other countries. The increasingly less fertilizer is used because valuable nourishment can be extracted from the animal excrement. Holland is no longer need to rely on importing phosphate. All these are the portrayal of the Dutch advanced agriculture.

The Dutch advanced agriculture can be traced back to the golden age. Since most of the food comes from Poland at that time, Dutch farmers are forced to focus on agricultural production. They try to apply science and technology to land cultivation and agriculture. The best example is the drainage of arable land and the improvement of soil fertility. These technologies later spread from the Netherlands to Europe and the world.

According to the World Trade Organization and the Netherlands Central Statistics Bureau, the total export of agricultural products in Holland in 2015 amounted to 81.3 billion Euro, ranking second in the world and second only to the United States. If agricultural-related products such as agricultural machinery and fertilizer are included, the total export can reach 90 billion Euro, accounting for 21% of the total annual export of the Netherlands. In such a small country of only 40,000 square kilometers, the agricultural products export ranks second in the world, which apparently does not rely on the local agricultural export. It is mainly due to the import, processing and re-export trade of the Dutch agriculture, which relies on the strong support of high-tech.

In Holland, when talking about the agricultural high-tech, people will think about Wageningen and the University of Wageningen. In addition to historical events, Wageningen is also known for the University of Wageningen which is dedicated to the instruction and research of life science. At the same time, Wageningen is the center of the Food Valley. The reason that Wageningen becomes one of the international cities in the Netherlands is largely due to the internationalization of the University of Wageningen. It is known as the Life Science City also because of the research of life science in the University of Wageningen.

The University of Wageningen (full name: The University of Wageningen and Research Center, in short: UR, WUR) is one of fourteen universities in the Netherlands, ranking at the top of the Dutch Higher Education Guide for nine consecutive years. It is a leading university in the field of life science in Europe and the world. WUR is the only university directly funded by the Dutch Ministry of Economic Affairs, and other universities are funded by the Dutch Ministry of Education.

WUR covers many professional curriculums from animal and plant science to food environmental science, and also includes non-natural science courses such as landscape design, economy and development research. Among them, the most prestigious are the plant science, environmental science and food science. Disciplines such as landscape design and international development are also very good. In general, WUR is a unique research type university.

WUR consists of the University of Wageningen, Marriott Lawrence Stannong Agricultural College and Agricultural Research Institution. It was founded in 1876, the earliest college providing agricultural education in the Netherlands. In the following four decades, it continued to grow and develop, and in 1918 formally approved by the Dutch government, it became an academic university, and changed its name to the Netherlands National Agricultural University. In 1997, it was merged with the scientific research institution under the Netherlands DLO Agricultural Research Foundation, known as the University of Wageningen and Research Center. "To explore the potential of nature to improve the quality of life" is the educational and research purpose of WUR. Over the past century, WUR has been committed to the higher education and scientific research of life science and natural resources related field, enjoying a high reputation and prestige in the international community.

WUR has launched and participated in 31 EU projects (80 in total) in terms of food safety and food quality, fully demonstrating its strength and international status in this field. It ranks third in the field of agricultural literature references in the world and fifth in the field of plant and animal science literature references. In terms of the agricultural technology, WUR has designed different greenhouse systems in different parts of the world, and combined efficient economic production with the least energy, water and nutrient inputs, to increase the productivity of agriculture around the world. Meanwhile, its unique biological control technology, that is using the natural enemies to deal with plant diseases and insect pests, expands the new direction in the research field of food safety and environment protection. WUR also makes a remarkable contribution in the field of environment science.

Wageningen Agro Business and Science Park gathers various small and medium-sized companies and research institutions in the fields of life sciences, food and health, with more than 8,000 scientific research staff. Under the research support of WUR, this park is known as the European life science center.

The Food Valley is the homes of the Dutch Food and Nutrition Research Cluster, which is comprised of many food and nutrition research institutes and companies representing the current state of advanced technology. It is considered as the largest food and nutrition development hub in the world today.

Since 1972, WUR provides a copy of agricultural economic report each year to the Dutch government. The report is about the economic situation of agriculture in the Netherlands, including the world's overall agricultural economy and political development, EU agriculture policy changes, Dutch agricultural complex and food industry development, changes in the rural areas of the Netherlands and environmental issues, descriptions of agricultural production structures and production factors, as well as the profitability of agricultural-related sub-sectors, as to provide the agricultural policy guidance to the Dutch government.

Obviously, the scientific research advantage is the core feature of the University of Wageningen, bringing Wageningen the unique international image. On this basis, the town establishes the leading industry focusing on the life science by relying on the platform of Agro Business and Science Park and Food Valley, and successfully attracts leading companies from all over the world to Wageningen. The University of Wageningen provides high quality graduates and the latest technology enterprises to those companies. This mutual dependent and mutual beneficial win-win relationship is beyond the traditional sense of investment, which is a perfect combination between the knowledge and market experience. The presence of companies also provides the university graduates more and better jobs. The Dutch National Animal and Plant Inspection and Quarantine Bureau is also located in the town.

In order to enhance the information exchange and technical cooperation between entrepreneurs, a meeting will be held once a month in the town's most famous Hotel de wereld, which provides a good entrepreneurial environment for those young entrepreneurs cherishing the dream to settle in Wageningen.

The cultural life in Wageningen is extremely rich. The town is highly internationalized, and overseas students can be seen everywhere. University buildings gather around the town and student apartments are in the city. Y-type student apartment, also known as the star apartment, becomes a unique landscape in Wageningen due to its shape and location.

Wageningen is not only a historic city, but also a city of life science, full of vitality.

Groningen
A miniature metropolis

Groningen is the most vibrant town of the Netherlands outside the Randstad with a population of 200.000. It has been the undisputed regional capital town of the North Netherlands for centuries. As a historic center it incorporates functions for trade, administration, production and education. Groningen University is one of the 100 best universities in the world with an international student population of over 30.000, and has many Chinese amongst its students and academic staff. In its urban design and layout Groningen is exemplary in matching a classic urban layout with 20 century typologies like shopping malls, university quarters, a ring road in a walkable and bikeable setting that extends into the surrounding open landscapes.

Nowadays when visiting Groningen on a regular work day, one can feel the rhythm of the academic life in the historic streets. At lunchtime the town is brimming with young people and academics finding their way to eat, to meet and moving to the next lecture or practicum session. All the movement of pretty students is set in a historic décor around Vismarkt and Grote markt and guided along the monuments that have been there for centuries like the Martini toren, the Koren-beurs and the Academy building. At the end of the day the town gets very lively again with students and academic staffs going to their evening appointments in one of its many lively street and neighborhoods. Being the regional capital for centuries Groningen has all the facilities of larger towns in a small package. The town feels confident and self assured as a person who has experienced a lot but always knows how to adjust to the new circumstances.

One could call Groningen a miniature metropolis since it features all the services of modern life ranging from museums, office buildings and classic heritage to friendly malls and a large football city. Walking around Groningen one feels the streets seamlessly flowing from one into another. There is always another lovely canal or park ahead. Old and new go hand in hand as matched pairs. Refined glass architectures are standing next to classic buildings. The friendly scale of everything provides a sense of comfort that stretches from the central squares to the meadows of the surrounding open land bordering the suburbs of the town.

The history of Groningen looms large when all the new students celebrate the local festival called "Bommen berend" at the beginning of the academic year. It is to celebrate the liberation of the town in 1695 when the bishop of Munster (in Germany) called Berend was the first to use cannons and bombs when trying to break the resistance of the northern capital of the seventh province of the Dutch republic. Groningen did not break and grew into a resilient town looking over the Northern provinces with a control and offering a hotspot for trade administration and education.

The university which was founded in 1614 started out as a school for clergyman. This was a common point of origin since religion played a central role in society and theological discussion was a major influence on government and administration. Already in the seventeenth century large parts of the students came from abroad, adding to an open and intellectual environment, stimulating discussion and innovation. This culture has developed into the global top 100 university, into the academic powerhouse that it is today. The main administration building called the academy building is still centrally located in the heart of the city. Other facilities like the ceremonial academy building, the university board, and the guesthouses are all located in the old town center. The impressive academic hospital complex is set on the edge of the old town built after the removal of the urban defense walls. North of the old town is Zernike park where the science and economy faculties are gathered. All buildings are within the bike distance of each other. The university is fully integrated in the town and mixed with neighborhoods and streets. The urban integration of the faculties and the students is the key to the lively and friendly atmosphere of Groningen. The intermingling of the university and town is an impressive feature, given the fact that most international universities (and Groningen's competitors for attracting students worldwide) are nowadays isolated in campuses outside the town and city centers.

Walking around Groningen and seeing the diligent buildings, one can't stop wondering where the people found the time and the money to create all these monumental buildings that are celebrations of civic accomplishment. Take for instance the Koren beurs, a top

100 monument in neoclassical style next to the vismarkt in the old town. In the time of its construction it was considered cutting edge with its mixture of classical symmetrical facades and refined steel constructions on the inside. It served as an epicenter of the trade in grain which was an important source of wealth for the city and its surroundings. The salty clay north of Groningen was a fertile ground for grain and other agriculture products that served to feed large parts of the Netherlands for centuries.

Another source of wealth was Groningens's monopoly on the extraction of peat in the southeast area of the town. This area known as the "Veen kolonien" (peat colonies) was an exploitation area connected by canals to the city of Groningen. The peat was used as fuel in the city and exported to the other parts of the Netherlands. The geographic imprint of the peat-economy in the region is still visible everywhere. The rich merchant's houses along the canals of Groningen are paid with peat money. Many of Groningen's waterways like the "Stadkanaal" were excavated for transportation of peat. The landscape of the veen-kolonien (nowadays a major tourism destination) near Groningen has been heavily influenced by the peat harvesting. After excavation of the turf the land would be used for agriculture and crops. The land was subsequently leased out to the farmers which contributed significantly to the city's finances. Many of the industrial infrastructures of the past have become the characteristic areas of Groningen nowadays.

Groningen counts many small scale walkable neighborhoods around the town center with low rise buildings and incidentally education or institutional buildings of a bigger size. A magic neighborhood is the Hortus kwartier located north of the Grote markt. The Hortus area was named after the academic gardens (hortus botanicus) in this area. Its streets and buildings are laid out in a seventeenth century plan with small streets and deep blocks. The small-scale parcelations with organic developments of individual houses and buildings are conceived in the pre-car area when people moved around on boats carriages and later the train. This charming atmosphere of small scales and walk ability of the pre-car area proves evermore popular today. One of the special residential typologies in this neighborhood is the old fashioned yard house where numbers of dwellings are positioned around a secluded garden with a gate to the street. In ancient times these courtyard houses were built to house widows or single women in a group with a private outdoor space. Nowadays these houses are also very popular with students and young couples who prefer a quiet living near the old downtown.

Walking around Groningen one can feel the layers of history that unfolded in the different areas of histories. Some of the town's most wonderful streets like the Ubbo Emmius Singel, the Heresingel and the Noorderplantsoen have a distinctly different character than the downtown. This late 19th century development was especially focused on capitalist bourgeoisie and produced numerous houses of standing and beauty. These developments are positioned on the edge of the town center where the urban defense walls were once positioned. In the late 19th century the government of the Netherlands decided that the defense of the country was national matter with a national infrastructure, hence the local defense walls in all Dutch towns were allowed to be removed based on the influential defense law of 1874. The removal of the ring of urban defense works and fortifications made way for modernization and expansion of the city in the 20th century.

Groningen as a city has been famous for an experimental and refined approach to its urban planning strategies. New typologies of the 20th century have been carefully added on an appropriate scale and integrated with the more classical parts of the city. This approach helped to maintain walk ability and a continuous mixture of larger and smaller buildings so the "organic town feeling" did not get lost in the 20th century. Even in the historic town center contemporary architecture goes hand in hand with classics of the past. Groningen has a long history in applying new typologies for residential compounds and transformation areas and has promoted engaging famous architects and avant-garde designers.

One example of the cities' architectural endeavors is, for instance, the Brinkflats along the Verbindingskanaal, where the famous Dutch architect Rem Koolhaas realized his first master plan and apartment buildings in 1988. Another famous example is the Waagstraat complex along the Groote Markt where the Italian architect Adolfo Natalini was chosen after a competition with a people's jury. Also the people's library designed by the Italian architect Giorgio Grassi at the Oude Boteringe straat is a wondrous humanistic building integrating classic shapes with modern architecture.

Maybe one of the most significant works of modern architecture is the Groninger museum. Positioned on islands in the verbindingskanaal, the museum is an important node on the pedestrian route from the train station to the town center. The museum is composed of three pavilions to keep its appearance small and transparent. Each pavilion is designed by another architect under supervision of the Italian architect Allesandero Mendini. The work is one eminent example of post modern architecture, a style that originated in the 80's. In general one can see this museum with its ambitious programming and its exquisite placement as a confirmation of the renewed popularity of the downtown after decades of decay. The process of downtown decay where people and companies left the downtown had influenced most western cities, after postwar waves of suburbanization.

Nowadays Groningen has a very international population to do its academic powerhouse. The town focuses more and more on the service industry and leisure entertainment relaying on its position as a regional center. The downtown is a hotspot for the more high standing activity and institutions. Its nearby leafy suburbs are excellently suited for growing families. Along the ring road the commercial and service facilities are gathered. The rural outskirts are perfect for relaxation and quiet life.

In 2019 the Groninger forum by NL Architects will be delivered. It's a downtown cultural service cluster under exceptional avant-garde architecture. The building includes libraries, exhibition centers, a cinema and entertainment venues in a vertical organization with a big view deck on the roof. The latest development will serve as a crown on the town center and an acknowledgement of Groningen as a globally oriented, 21st century, regionally positioned, science- leisure-service town.

后记 POSTSCRIPT

中国近40年的改革开放史，主线其实就是工业化、城市化、现代化的历史，并集中体现在城市化上。城市是政治经济、生产生活、交通交流的空间物质载体，在现代化形象塑造和生活改善需求的双重动力驱动下，看似无规律的"井喷"、跳跃的方式扩大的同时，也暗合着自然发展的内在逻辑：从"旧貌换新颜"简单的大拆大建，到"开发区""工业园""居住区"和"商业区"功能细化的城市分块组团，再到将功能单一的区块整合形成更大的"城市综合体"。这一过程，是热衷于塑造"大城市"的结果。在人口极速膨胀、GDP指标迅速攀升的同时，重蹈西方城市病之覆辙，也成为必然。

伴随着经济和生活需要的城市功能划分，不少富有生机的城市和居住区变得死气沉沉。为汽车而营建的发达道路网络，将城市文化和人之间的交流与联系，分隔得支离破碎。多个体量庞大、造型类似的建筑物拥挤在同一区域，让城市形象模糊而迷失方向。噪音、粉尘、雾霾、拥堵让居民们不堪承受。文化的丢失、景观的趋同，让同质化现象不断扩大和重现。敏锐的规划师早已对此表达了担忧，呼吁社会对满足区块功能和城市文化传承方面予以重视。然而，大多数规划师在人口膨胀快节奏压迫和功能主义规划要求中已疲于奔命，缺乏对城市的特征和内在精神展开深入的研究与表现，使城市趋同性和城市视觉意象的混乱与模糊问题愈加严重。城市毕竟是人居的，城市规划发展必须以人为本，而不是以城市为本。规划师要从经济、产业、社会、心理的视角去考虑城市的发展，必须平衡城市在经济、环境、社会与人之间的矛盾；要对在城市边界范围、空间尺度不断扩大后，在城市生活的人们，直接地观看、聆听与交往可能被减弱而引起的心理失衡，引起极大关注。

事实上，社会与过去对比，城市功能的巨大改善带来的喜悦已趋于平缓。相反，精神生活的缺失感开始出现，加之出国游赏，国外富有特色的城市，特别是文化气息浓厚的小镇带来的视觉冲击与心灵感受，让人们更加强烈地不满于"千城一面"的单调以及望不见山、看不到水、找不到乡愁的文化缺失。这种巨大的心理落差形成的负面情绪，正通过现代化的网络传媒，迅速在社会蔓延。然而，一个时尚现代、功能齐全、机遇多多的城市，依然在打动、吸引乡村居民。可是，潮水般地涌向城市，只能加剧城市病；无序的开发建设，便是继续让城市病泛滥。所以，国家层面的战略不断调整，从提出城市化，到改称城镇化，又提升为新型城镇化，包括美丽乡村建设。终于，到2016年，一个新兴实体，人类诗意栖居、激情工作、悠然畅游的新载体，如雨后春笋般在我国崛起，这就是特色小镇。"非城非乡非园区，宜居宜业宜旅游"的特色小镇，若用诗意的语言表述，就是：城市功能，田园风光。诗和远方，就在身边。

中国发展特色小镇的意义在于：

1.推进新型城镇化，也是美丽乡村的升级版；

2.疏解大城市压力；

3.有利于产业调整，尤其是发展新型产业（如云产业、设计中心、艺术、观光农业、旅游等）；

4.推动供给侧结构性改革，特别是个性化产品的研发、订制化生产；

5.特色小镇的建设，还为创业者、投资者带来新一轮的发展机遇。

与大城市相比，小镇的低房价极大地降低了创业成本，比如，浙江基金小镇，房价是上海陆家嘴金融CBD的二十分之一。各种新兴和新型产业，也创造了许多新的就业岗位，有志青年可以不再把眼光盯在北上广。而对投融资而言，仅国家级特色小镇就要建1000个，省级不少于3000个，有条件建设特色小镇的地方，全国应该是7000个以上。按平均投资50亿元计算，总投入将达到35万亿以上，投资拉动的战略上不再是"铁公鸡"。特色小镇的建设，还为许多机构与专业人员带来机遇，如规划设计、教育培训、文化、技术、管理的交流，以及生态城市、智慧城市、海绵城市等方面的新产品开发，此外，还有体育、健康、交通、养老等新产业的发展。特色小镇，是人类诗意栖居的康庄大道，是新型城市化和后工业时代产业的发展之道，是新时代观光、游学、创意、养生的新旅游之道，是各大机构与专业人员拓展投资渠道、搭建交流平台、开展咨询活动、实现"互联网+"之道。规划、建设、运营特色小镇，需要小镇及其相关方面共同协作，包括借鉴西方的理念、技术，因为西方已经走在前面，这是全人类的文明成果，我国可以借鉴发挥后发优势，实现弯道超车！

荷兰是一个神奇的国度，不仅在西方大国崛起、现代化商贸金融业上走在前列，而且在其山水之间，依然保存着花田、风车和静谧的村庄。一个个既有城市功能、又有田园风光，既有文化底蕴、又有旅游魅力的小镇星罗棋布，成为居住者的天堂、创业者的乐园、游客的梦乡。荷兰位于欧洲西部，东面与德国为邻，南接比利时，西、北濒临北海，地处莱茵河、马斯河和斯凯尔特河三角洲。荷兰1463年正式成为国家，现有国土总面积达41864平方千米，分设12个省，建国500多年，海堤、风车、郁金香和奶酪，成为最显著的特征和最难忘的形象，各具特色的小镇有几百个，其中约有30~50个的特色小镇举世闻名。

笔者在荷兰生活和工作多年，为了寻求特色小镇的设计灵感、汲取可用于中国小镇规划建设的营养，用时1年，走访了100多个小镇，选取其中有代表性的15个，娓娓道来，著成此书。笔者以为，我国地域辽阔，国土面积为荷兰的230倍，若以荷兰小镇的面积为尺度，我国可形成6.5万个小镇。从小镇内涵来看，中华民族五千多年来，在具有典型特色的56个民族，围绕在长江、黄河为代表的母亲河畔，以五岳为代表的山峦间，在横跨寒、温、热、亚热4个气候带中，以"天人合一"理念、"厚德载物"境界、"自强不息"精神，展开着生存、繁衍、创造和发展的生活，积淀了灿烂而丰富多彩的文化内涵，是小镇取之不尽的特色源泉。改革开放至今，我国已在荒芜的土地上建成像浦东、深圳这样的现代化、国际化大都市，那么，我国完全有信心、有能力、也有必要建成一批传承中国文化，彰显中国特色的美丽的、诗意的、惊艳世界的特色小镇。

历史的经验显示，能够接受先进理念、吸纳外来文化、运用最新科技，才是发展王道。中国文化的特征之一，便是包容开放。因此，在中国特色小镇的规划、建设、运营过程中，荷兰小镇无疑可成为无言之师。我国不仅要学习借鉴，还要思考创新，以青出于蓝，因为我国有960万平方公里的锦绣河山，有五千多年历史积淀的灿烂文化，更有13亿勤劳智慧的华夏儿女！

以上就是笔者创作本书的初衷，本书亦可作为前往荷兰考察、旅游的读本。笔者以游览者的口吻来叙述，有风光描写，有情感抒发，有时空感悟，还有中荷对比，并配有精美图片。因成书仓促，书中错误、不妥之处，在所难免，敬请指正！